IN THE
BALANCE
ALGEBRA LOGIC PUZZLES

Grades 4–6

Lou Kroner

Creative Publications®

Senior Editor: Glenda Stewart
Editor: John Lanyi
Designers: Gregg McGreevy, Meg Saint-Loubert
Production Coordinator: Kate Rapson
Production Services: Dovetail Publishing Services
Translation Services: Sybil de Munsuri
Cover Design: Al Golden

©1997 Creative Publications
Two Prudential Plaza, Suite 1175
Chicago, Il. 60601
Printed in the United States of America

ISBN: 0-7622-0551-2
4 5 6 7 8 9 03 02 01 00

CONTENTS

TO THE TEACHER

Mobiles have always fascinated me. As a child, I watched Alexander Calder's mobile "20 Leaves and an Apple" spin a leisurely course in the air currents at the Cincinnati Art Museum. But this book is not about Calder's creative genius or the delicate balance of the kinetic sculptor's art. The mobiles in this book were designed with math teachers and their pupils in mind.

These activities allow students to apply mathematical reasoning skills and problem-solving strategies and to develop number sense by exploring relationships between numbers. My own students have completed them independently and in cooperative groups.

The puzzles are arranged in an approximate easy-to-harder progression. Individual students will always find certain puzzles easier or harder than others. Sometimes, rather than being more difficult mathematically, a puzzle is more difficult because its complexity requires more sophisticated organization.

One suggestion to extend the thinking of these puzzles for highly able or motivated students is to have them design original puzzles. Creating a working puzzle and describing the step-by-step strategy required to solve it provides a creative challenge.

ACKNOWLEDGMENTS

As I worked on this project, I received lots of encouragement from family, friends, and colleagues.

Most of all, thanks to my family—Mary Kay, Katie, and Alan. Without their patience and support, this project would have faded away long ago.

Thanks to Sandy Lingo (J. F. Dulles School), Cheryl Tallman (Our Lady of Visitation School) and Todd Forman (Summit Country Day School) for their professional encouragement and feedback.

Lou Kroner
Cincinnati, Ohio

DIRECTIONS

In these puzzles, each shape stands for a number value or "weight." To solve these puzzles, you must make the weights in each part of the mobile balance from left to right, just as a sculptor must balance all the parts of a mobile.

Use your sense of number logic and your experience with number relationships to help you discover the value of each shape. As you work, remember these rules:

- The right and left sides of each horizontal beam must balance.

- Each shape has a unique and consistent weight within a puzzle, and no shapes weigh zero.

- There are no "useless" clues. For example, if a clue says that the square's weight is a multiple of the triangle's weight, you may safely assume that the triangle does not weigh one.

- All weights are either one- or two-digit, positive, whole numbers.

- A piece hanging directly below the fulcrum *does not* affect the balance between the left and right arms. Although this piece has its own definite weight, it might be considered "neutral" for balancing the other two arms. (See sample 3 on the next page.)

- Size of pieces has no relation to weight.

- These mobiles are exercises in balancing number values. They do not take into account distance from the fulcrum or any other principles of physical science.

Sample 1

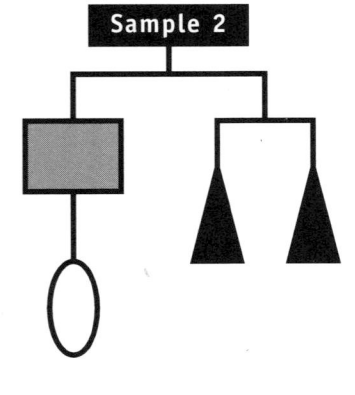

Total = 16

Sample 2

Total = 24

Clue: ▲ − ◯ = 2

Sample 3

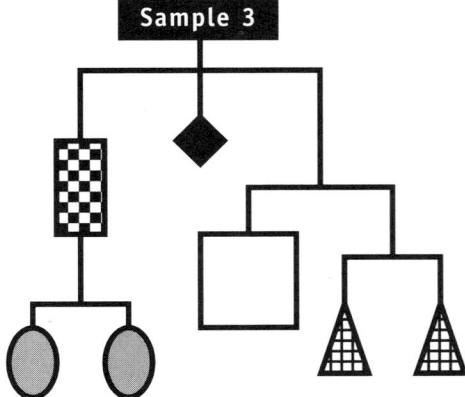

Total = 31

Clues: ▦ − 1 = ◆   4 × ⬭ = ▦

Solution

Since the whole of the left side is taken up by the circle, it must weigh 8. Since half of the right side, which also weighs 8, is taken up by the triangle, its weight must be 4. To balance out this 4, each of the rectangles must weigh 2.

Solution

Each half weighs half of the 24, or 12. Therefore, each of the triangles must weigh 6. That also means that there could be the following combinations for the rectangle and ellipse on the left side: 7 + 5, 8 + 4, 9 + 3, 10 + 2, 11 + 1. Of these, only one combination, 8 + 4, fits the clue given. Therefore, the ellipse must weigh 4 and the rectangle must weigh 8.

Solution

The problem here is that, because of the diamond, we don't readily know the values of the two arms. We do know that each arm can, at the most, weigh 15. But 15 cannot be halved to suit the right arm, so we know that each of the arms must weigh 12, 8, or 4. (Since there are three shapes on each side, we can eliminate 2.)

On the left arm, the clue tells us that it takes 4 ellipses to equal a rectangle. If the ellipse weighed 3, the rectangle would weigh 12, giving a total weight of 18 for the left arm, three more than the maximum. That means that the ellipse must weigh 2 and the rectangle 8. With this arm totaling 12, weights for the square (6), triangle (3), and diamond (7) are forced.

If-Then Table

Although each person's strategy for solving these problems will differ, one that I have found helpful is the If-Then table. Using this simple strategy, students prepare a table that lists all the possible combinations of weights. This table helps students organize their work, maintains a record of their trials, and presents the weights in a visual format. Here is a sample:

Total = 20

Clue: ◯ + ■ < 14

■	◯
18	1
16	2
14	3
12	4
10	5
8	6
6	7
4	8
2	9

Using the clue provided, students can now eliminate all but three possible combinations. Of course, a multi-columned table may be necessary or helpful for more complex problems and relationships.

Even or Odd Number?

Another important strategy involves learning whether a shape represents an even or an odd number. Discovering (or rediscovering) relationships like "an even number subtracted from an even number yields an even number" will help as a pupil analyzes a problem and develops a strategy.

Two Final Notes

1. As mentioned above, there are several—even many—right ways to solve these puzzles. The solutions that follow are only suggestions or possibilities. An attempt has been made to include several types of strategies, so you'll notice that some of the solutions seem more straightforward while others seem more involved. Sometimes, using a table instead of putting solutions into word form makes a solution seem much easier.

2. To identify shapes from the mobiles in these solutions, the following convention has been used: R indicates right arms, L indicates left arms, and C indicates center arms. To get to a shape labeled RRRL, you would take the right arm from the top, then the right arm of that arm, then the right arm of that arm, and then (finally!) the left arm of that arm.

Name _____

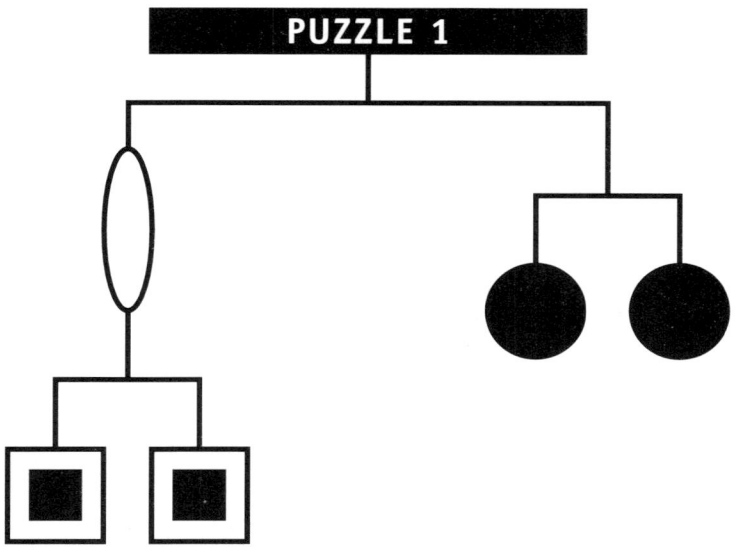

PUZZLE 1

Discover the value of each of the shapes. The total
weight is 36. All shapes weigh less than ten.
Additional clue:

4

Solution

1. By halving the weights of the arms, you learn the value of the circles (9 each) in RL and RR.

2. Ellipse must be an even number (when subtracted from an even number, 18, it yields a number that can be halved). Knowing that all shapes weigh less than ten, you can use an If-Then table to list three possibilities for ellipse:square: 2:8, 4:7, and 8:5. But only the third possible combination fits the second clue, so you force square (5) and ellipse (8).

Values

● = 9   ■ = 5   () = 8

Key	Sample Sequence
R = right arm **L** = left arm **C** = center arm	RRRL = (from the top) right arm, right arm of that arm, right arm of that arm, left arm of that arm.

Name _____

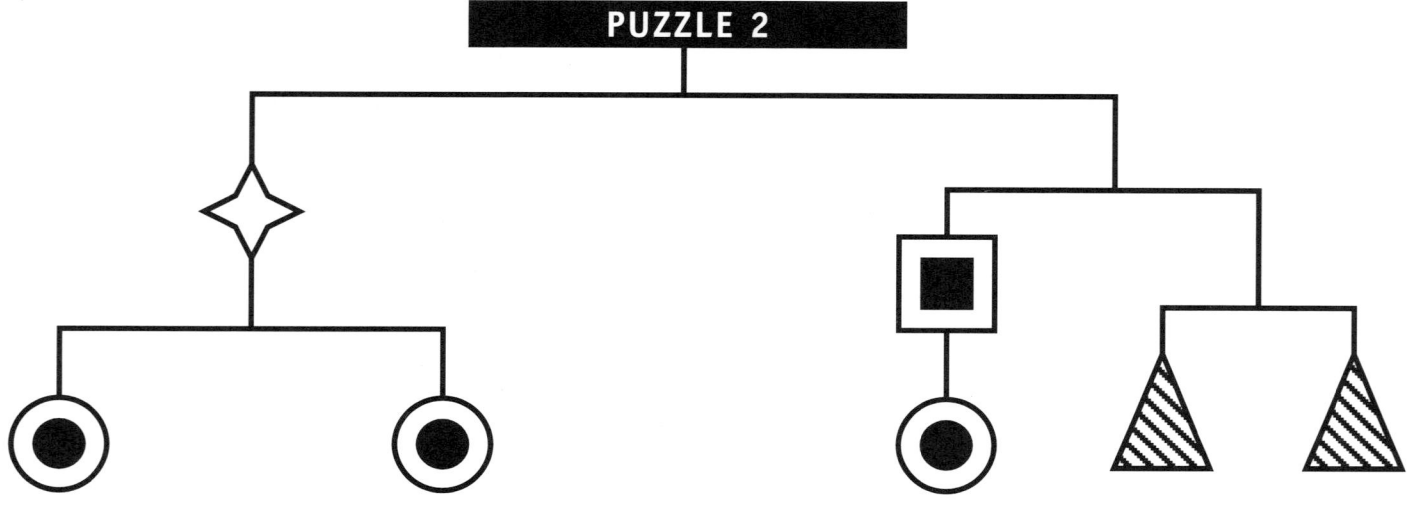

Discover the value of each of the shapes.
The total weight is 32. Clue:

$$\diamond - 2 = \blacksquare + \odot$$

Solution

1. Breaking down the weights of the two arms, you learn that triangle equals 4.

2. Using an If-Then table to list possible square:circle combinations (at RL), you find 1:7, 2:6, 3:5, 5:3, 6:2, and 7:1.

3. Using an If-Then table to list possible circle:star combinations (at L), you find 7:2, 6:4, 5:6, 3:10, 2:12, and 1:14.

4. Combining 2 and 3 from above, you list square:circle:star combinations and find 1:7:2, 2:6:4, 3:5:6, 5:3:10, 6:2:12, and 7:1:14. The only combination that fits the clue given is 5:3:10, so you know square (5), circle (3), and star (10).

OR

2. Given step 1, square plus circle equals 8. Given the clue, star is forced (10).

3. Knowing star forces circle (3).

4. Knowing circle forces square (5).

Values

$\triangle$ = 4 $\blacksquare$ = 5 $\bullet$ = 3 $\diamondsuit$ = 10

Key	Sample Sequence
R = right arm **L** = left arm **C** = center arm	RRRL = (from the top) right arm, right arm of that arm, right arm of that arm, left arm of that arm.

Name _____

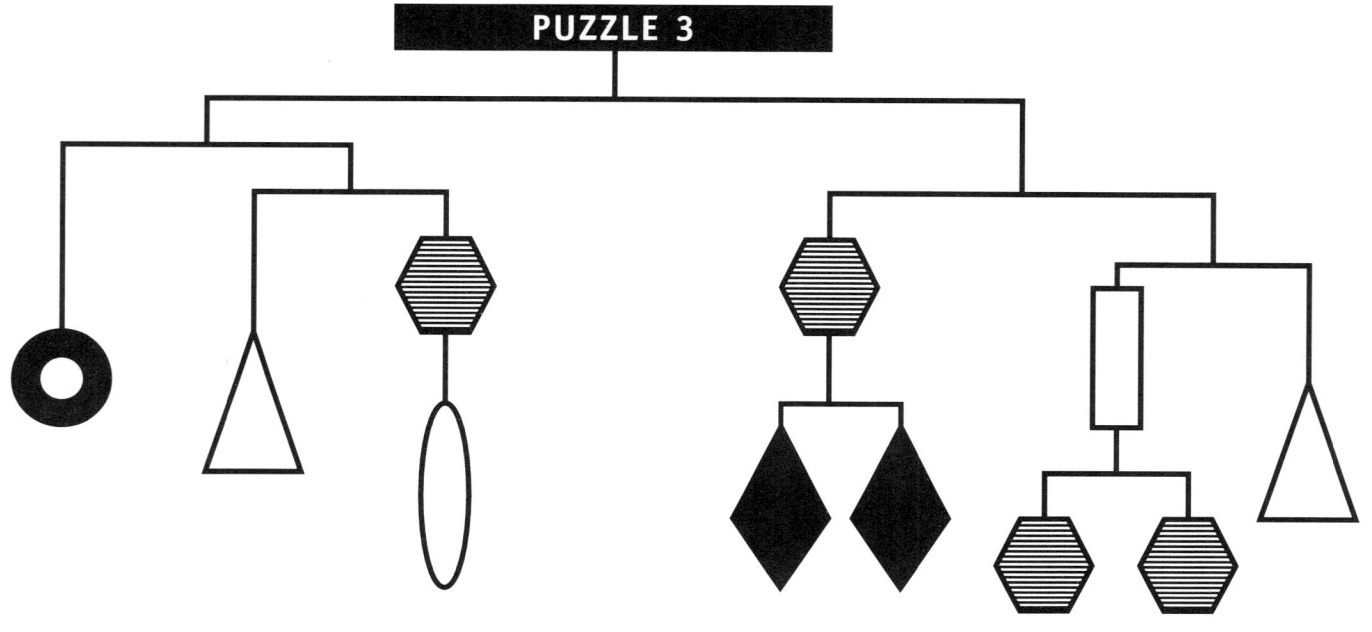

Discover the value of each of the shapes.
The total weight is 40.

IN THE BALANCE   Grades 4–6

Solution

1. By halving the weights of the arms, you can learn the values of the circle (10) in LL and the triangle (5) LRL.

2. Since the rectangle plus the two hexagons in RRL weigh 5 together, you can list in an If-Then table two sets of possibilities: 3 + 1 + 1 or 1 + 2 + 2.

3. But the hexagon appears with two diamonds for a value of 10 in RL. This means that the hexagon must have an even-numbered value, or 2. This forces the rectangle (1), the diamonds (4 each), and the ellipse (3).

Values

$\bigcirc$ = 10 $\triangle$ = 5 $\hexagon$ = 2 $\square$ = 1

$\blacklozenge$ = 4 $\oval$ = 3

Key	Sample Sequence
R = right arm **L** = left arm **C** = center arm	RRRL = (from the top) right arm, right arm of that arm, right arm of that arm, left arm of that arm.

PUZZLE 4

Discover the value of each of the shapes.
The total weight is 96.

10

Solution

1. Breaking down the weights of the arms into halves, you learn diamond (6) at LRLR and circle (3) at LRLL.

2. You know that square must be even (when subtracted from an even number, 48, it yields an even number) and less than 12 (there is a square at LRR).

3. Using an If-Then table to list possible square:pentagon:ellipse combinations, you find only 8:10:5 yields all integer weights. (Square equals 4, for example, yields pentagon equals 11 and ellipse equals 5.5!) Square therefore equals 8.

4. Knowing square forces rectangle (4) at LRR.

5. Knowing rectangle forces hexagon (20) at LL.

Values

◆ = 6   ● = 3   ▦ = 8   ⬡ = 10

◍ = 5   ▯ = 4   ⬢ = 20

Key	Sample Sequence
R = right arm **L** = left arm **C** = center arm	RRRL = (from the top) right arm, right arm of that arm, right arm of that arm, left arm of that arm.

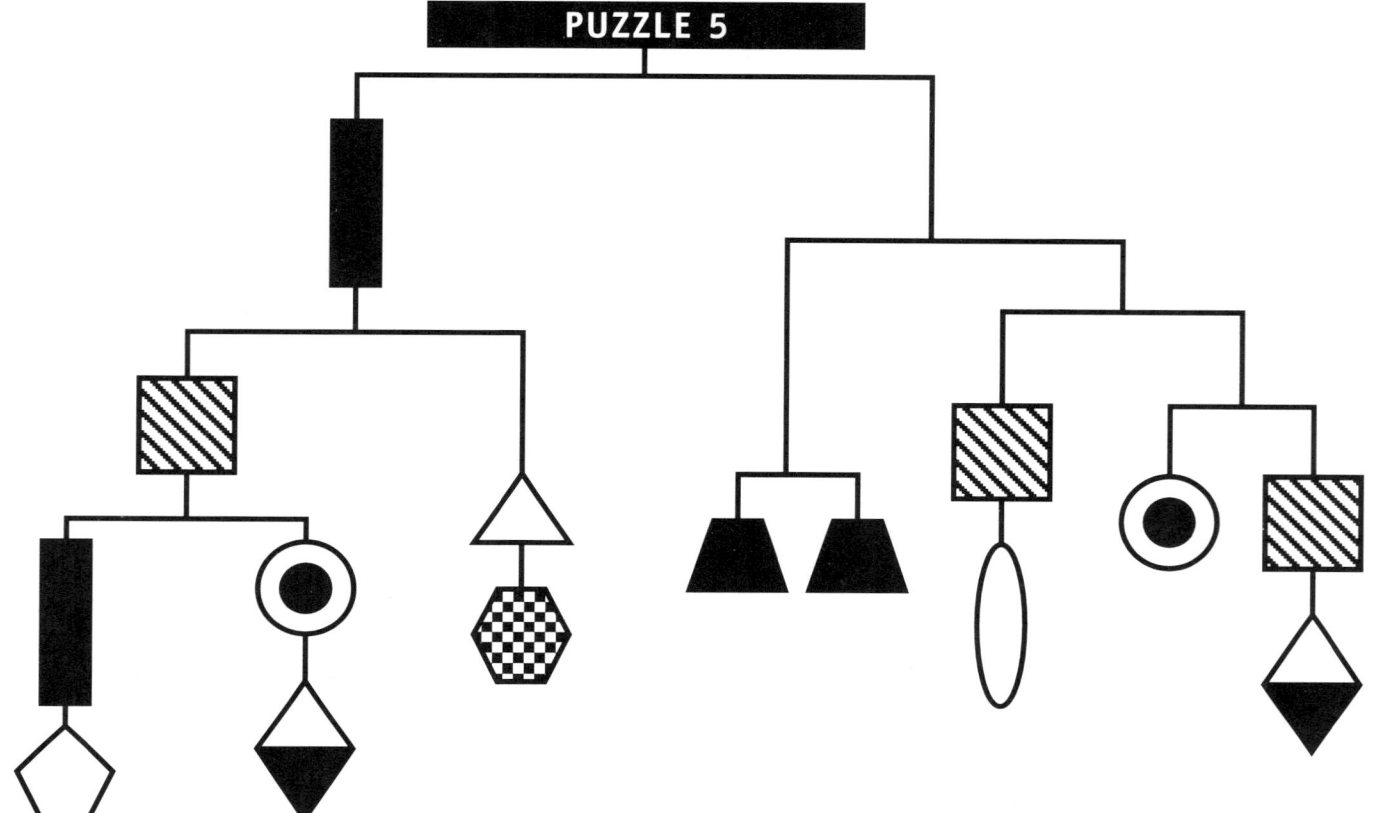

PUZZLE 5

Discover the value of each of the shapes. The total weight is 80. Only one shape weighs more than nine. Additional clues:

◆ + 1 = ▨   △ < ⬡

Solution

1. By halving the weights of the arms, you learn the weights of the trapezoids (10 each) in RL and the circle (5) in RRRL.

2. On RRRR, you see that the square and diamond weigh 5 together. The second clue given tells you that the weight of the square is one more than the weight of the diamond, forcing square (3) and diamond (2).

3. Step 2 in turn forces the ellipse (7) in RRL.

4. Knowing the weights of the circle and diamond on LLR, you learn the weights of LLL (7) and LR (17). Knowing that LL + LR + rectangle = 40, you learn rectangle (6).

5. Knowing the rectangle forces the pentagon (1) in LLL.

6. The third clue tells us that the triangle on LR weighs less than the hexagon. Their total weight is 17. Eliminating weights already used leaves us with only two possibilities: 4 + 13 or 8 + 9. Since only one shape weighs more than 9 and trapezoid equals 10, the weights of the triangle (8) and hexagon (9) are forced.

Values

▲ = 10   ◉ = 5   ◩ = 3   ◆ = 2

◯ = 7   ▮ = 6   ⬠ = 1   △ = 8

⬡ = 9

Key	Sample Sequence
R = right arm **L** = left arm **C** = center arm	RRRL = (from the top) right arm, right arm of that arm, right arm of that arm, left arm of that arm.

PUZZLE 6

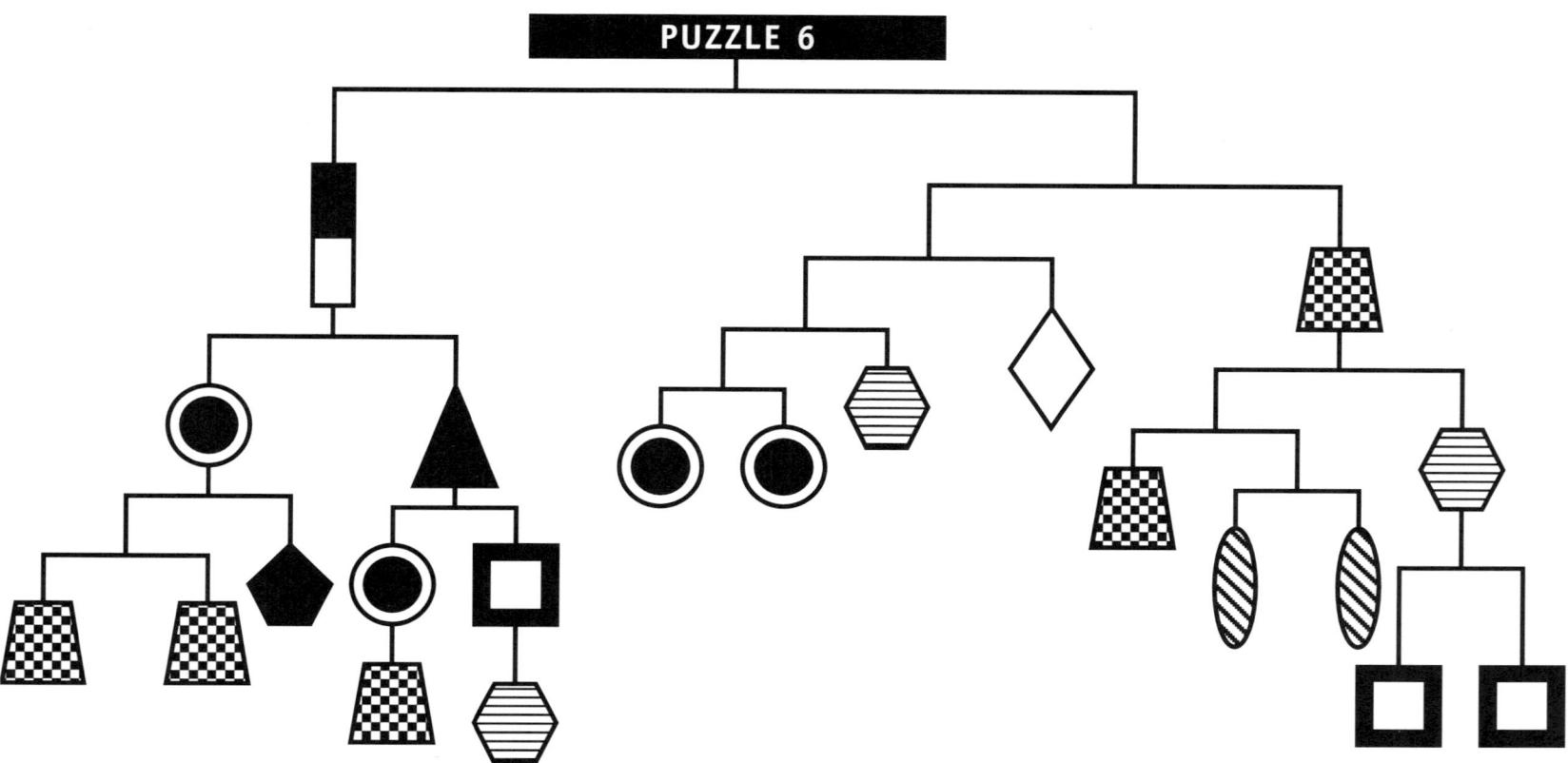

Discover the value of each of the shapes.
The total weight is 160.

IN THE BALANCE Grades 4–6

Solution

1. By halving the weights on the arms, you learn circle (5), hexagon (10), and diamond (20).

2. At RRLR, you could substitute a trapezoid for the two ellipses (since they balance with the trapezoid at RRLRL, you know they're equal in weight). This would make RRL equal 2 trapezoids. Since RRR balances with RRL (2 trapezoids), you could substitute 2 trapezoids for the hexagon and squares. This gives you RR (40) equals 5 trapezoids, or trapezoid (8).

3. Knowing trapezoid forces ellipse (4) at RRLRR and RRLRL and square (3) at RRR.

4. Knowing trapezoid also forces pentagon (16) at LLR.

5. Knowing pentagon and trapezoid forces triangle (11) at LR.

6. Knowing the above forces rectangle (6) at L.

Values

⊙ = 5   ⬡ = 10   ◇ = 20   ▦ = 8

⬭ = 4   ◻ = 3   ⬟ = 16   ▲ = 11

▯ = 6

Key	Sample Sequence
R = right arm **L** = left arm **C** = center arm	RRRL = (from the top) right arm, right arm of that arm, right arm of that arm, left arm of that arm.

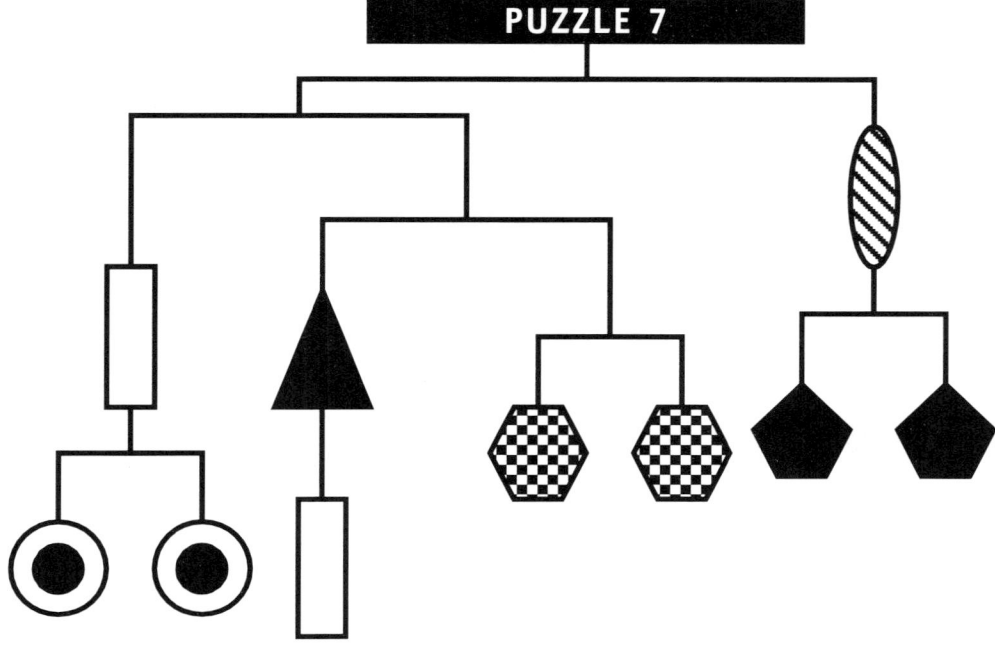

Discover the value of each of the shapes.
The total weight is 48. Clues:

Solution

1. Halving the weights of the arms, you learn hexagon (3).

2. From LL, you know that rectangle is even (subtracted from an even number, 12, it yields an even number). Listing in an If-Then table possible combinations of rectangle:circle, you get 10:1, 8:2, 6:3, 4:4, 2:5. Only one combination fits the first clue given, circle is greater than rectangle, so you learn circle (5) and rectangle (2).

3. Knowing rectangle forces triangle (4) in LRL.

4. The second clue tells you that pentagon is greater than the sum of circle (5) and hexagon (3), or 8. Making a list of possibilities for ellipse and pentagon, there are only three pentagon values that fit this clue: 9, 10, 11. A pentagon value of 10 would force ellipse to 4, and a pentagon value of 11 would force the ellipse to 2. Since both of those possible ellipse values are already assigned, you eliminate them, leaving a pentagon value of 9 and ellipse value of 6.

Values

⬢ = 3   ◉ = 5   ▯ = 2   ▲ = 4

⬟ = 9   ⬮ = 6

Key	Sample Sequence
R = right arm **L** = left arm **C** = center arm	RRRL = (from the top) right arm, right arm of that arm, right arm of that arm, left arm of that arm.

Name _____

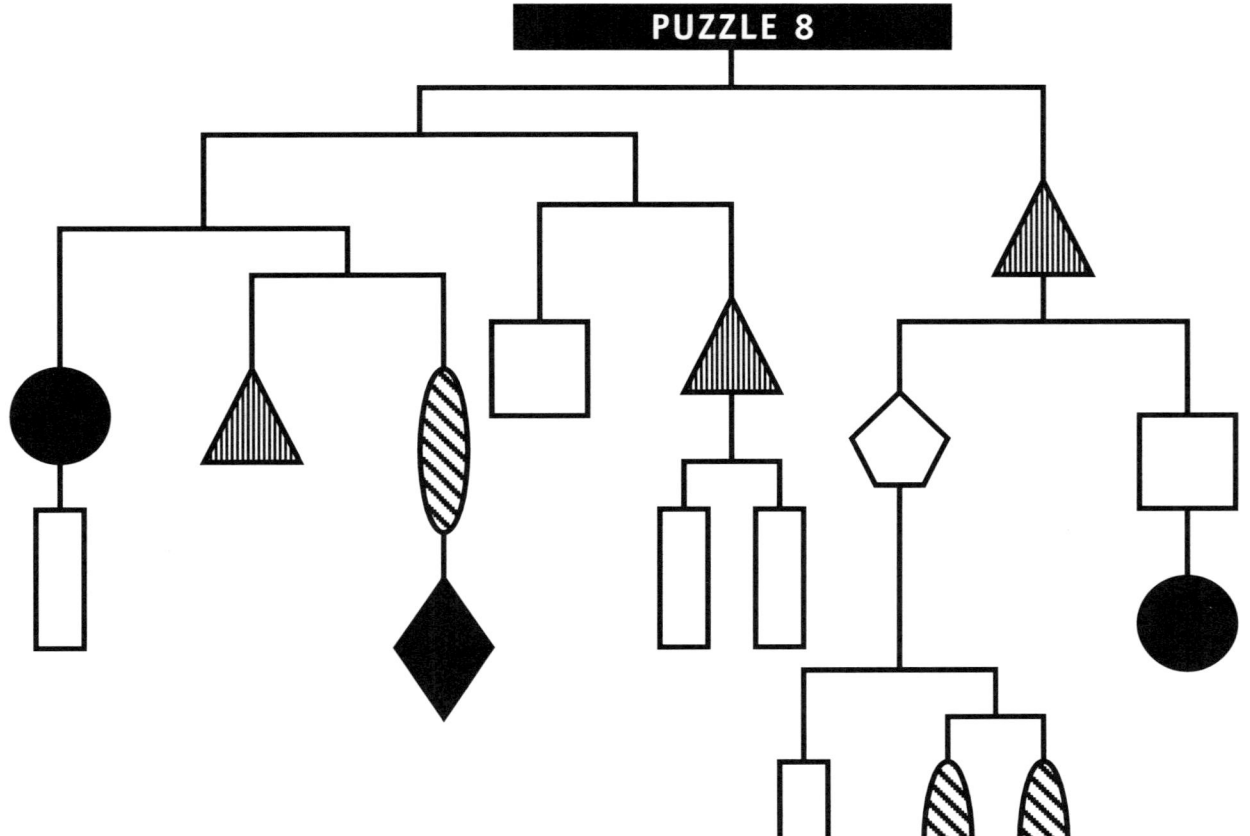

PUZZLE 8

Discover the value of each of the shapes.
The total weight is 64. Clue:

⬭ < ◆

Solution

1. Breaking down the weight of the arms into halves, you learn the weights of triangle (4) at LLRL and square (8) at LRL.

2. Given the clue, the only possibility at LLRR must be ellipse (1) and diamond (3).

3. Knowing triangle forces rectangle (2) in LRR.

4. Knowing rectangle and ellipse forces pentagon (10) in RL, and knowing pentagon forces circle (6) in RR.

Values

= 4   = 8   = 1   = 3

= 2   = 10   = 6

Key	Sample Sequence
R = right arm **L** = left arm **C** = center arm	RRRL = (from the top) right arm, right arm of that arm, right arm of that arm, left arm of that arm.

PUZZLE 9

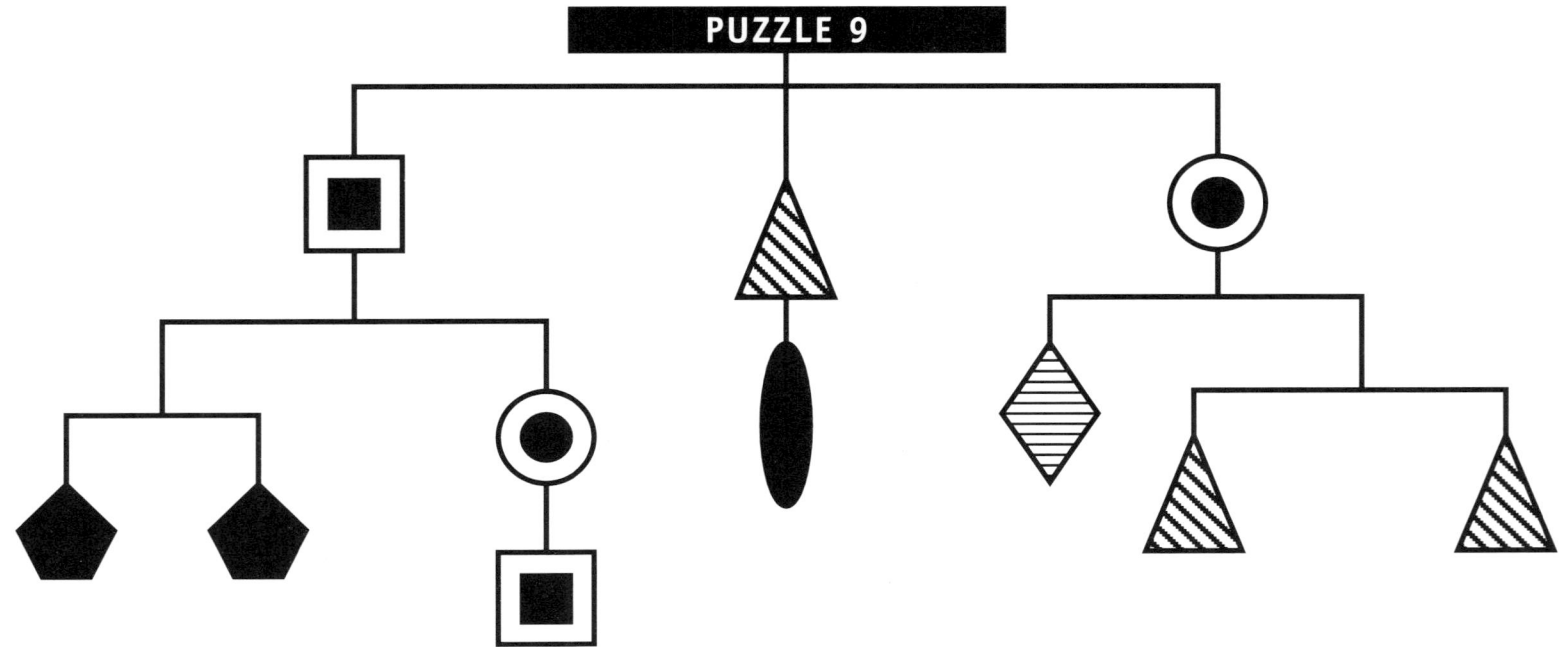

Discover the value of each of the shapes. The total weight is 129. Each of the arms is equal in weight. Additional clue:

⬭ = 5 ◼

Solution

1. Square (at L) and circle (at R) must be odd because, when subtracted from 43, they yield numbers that can be halved.

2. Using an If-Then table to list possible square:ellipse combinations that fit the second clue, you learn 1:5, 3:15, 5:25, and 7:35.

3. Using the above ellipse weights to list possible ellipse:triangle combinations, you learn 5:38, 15:28, 25:18, and 35:8. Of these, only 35:8 allows the triangle to work at RRR and RRL, so you know triangle (8), ellipse (35), and square (7).

4. Knowing triangle forces diamond (16) at RL.

5. Knowing square forces pentagon (9) at LLL and LLR.

6. Knowing the above forces circle (11) at LR.

Values

$\triangle$ = 8 $\mathbf{\varnothing}$ = 35   ■ = 7 $\Leftrightarrow$ = 16

⬠ = 9   ◉ = 11

Key	Sample Sequence
R = right arm **L** = left arm **C** = center arm	RRRL = (from the top) right arm, right arm of that arm, right arm of that arm, left arm of that arm.

Name _____

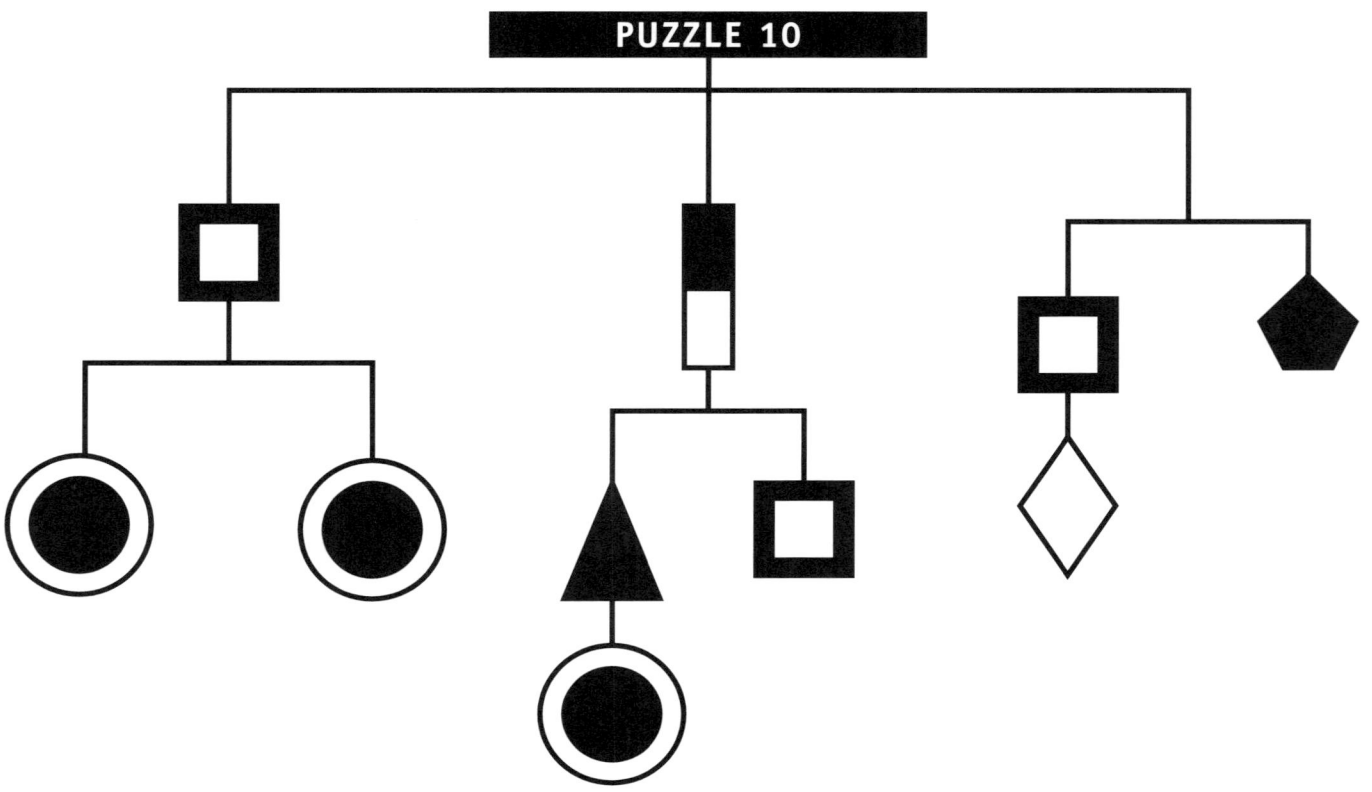

Discover the value of each of the shapes. The total
weight is 54. The three arms are equal in weight.

Solution

1. By halving the weights of the arms, you learn pentagon (9) at RR.

2. Using an If-Then table to list possible square:circle combinations that will work at L, you find 2:8, 4:7, 8:5. (6:6 is eliminated because of the duplicate weights.)

3. Building on what you learned in step 2, you list square:circle:diamond combinations and find 2:8:4, 4:7:3, and 8:5:1. Of these three, only 8:5:1 allows square and diamond to balance with pentagon at RR. So you learn square (8), circle (5), and diamond (1).

4. Knowing the above forces triangle (3) at CL and then rectangle (2) at C.

Values

⬠ = 9   ◼ = 8   ◉ = 5   ◇ = 1

▲ = 3   ▮ = 2

Key	Sample Sequence
R = right arm L = left arm C = center arm	RRRL = (from the top) right arm, right arm of that arm, right arm of that arm, left arm of that arm.

PUZZLE 11

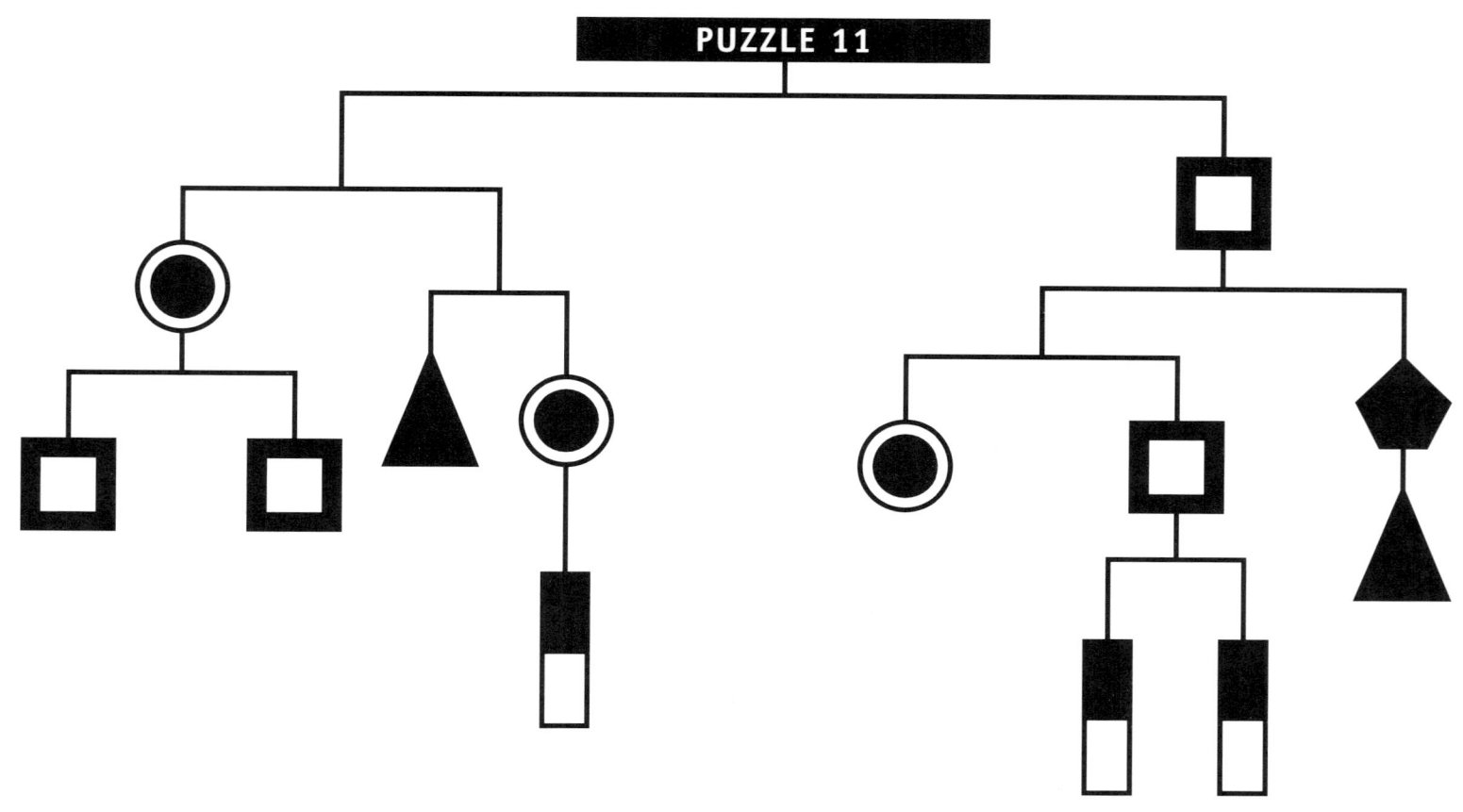

Discover the value of each of the shapes.
The total weight is 56.

IN THE BALANCE Grades 4–6

Solution

1. By halving the weights of the arms, you learn triangle (7) at LRL.

2. Circle at LL must be even because, when subtracted from 14, it yields an even difference.

3. Using an If-Then table to list possible circle:rectangle combinations (that equal 7 as at LRR), you find 2:5, 4:3, and 6:1.

4. Using in an If-Then table to list possible circle:rectangle:square combinations (at RL), you find 2:5:6, 4:3:5, and 6:1:4. But square values of 6 or 5 would be too much at RL, so you learn circle (6), rectangle (1), and square (4).

5. Knowing the above, you learn pentagon (5) at RR.

Values

▲ = 7   ◉ = 6   ▯ = 1   ◻ = 4

⬟ = 5

Key	Sample Sequence
R = right arm **L** = left arm **C** = center arm	RRRL = (from the top) right arm, right arm of that arm, right arm of that arm, left arm of that arm.

PUZZLE 12

Discover the value of each of the shapes.
The total weight is 50. Clue:

26

Solution

1. Square and triangle must be odd because, when subtracted from an odd weight (25), they each yield an even weight.

2. Using an If-Then table to list possible square:ellipse combinations, you find 1:12, 3:11, 5:10, 7:9, 9:8, 11:7, 13:6, 15:5, 17:4, . . . Of these, only 1:12, 9:8, and 17:4 allow for ellipse weights that can be subdivided twice to equal circle (at RRL) and triangle values (at RRR).

3. Taking into consideration that triangle must be odd (step 1), list two possible square:ellipse:circle:triangle combinations: 1:12:6:3 and 17:4:2:1. But the clue given eliminates the second of these, so you learn square (1), ellipse (12), circle (6), and triangle (3).

4. Knowing triangle (at L) forces rectangle (11) at LL.

5. Knowing rectangle (at LL) forces pentagon (5) at LR.

Values

▤ = 1   ⬮ = 12   ⬤ = 6   ▲ = 3

▯ = 11   ◈ = 5

Key	Sample Sequence
R = right arm **L** = left arm **C** = center arm	RRRL = (from the top) right arm, right arm of that arm, right arm of that arm, left arm of that arm.

PUZZLE 13

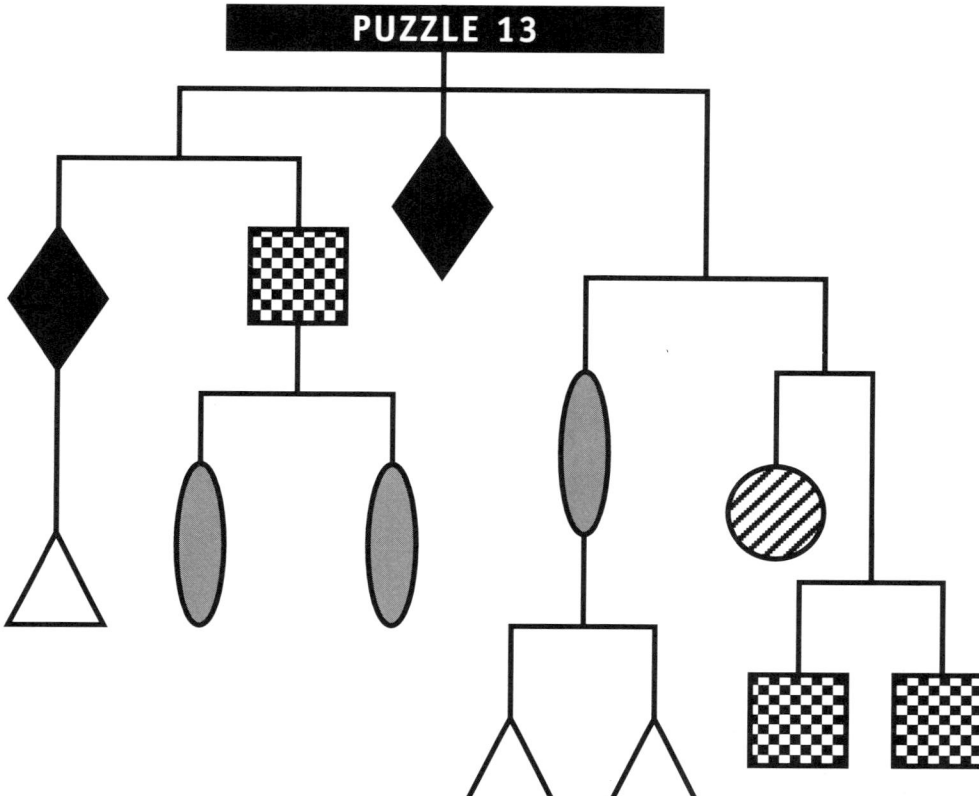

Discover the value of each of the shapes.
The total weight is 75.

Solution

1. Diamond must be odd, because when subtracted from 75, it yields an even number.

2. After the original split, R is halved three more times.

3. Using an If-Then table to list possible diamond:right arm (R) combinations (1:37, 3:36, 5:35, . . .), you find only two combinations in which R can be halved three times: 11:32 and 27:24.

4. If diamond weighs 27, then L and R must weigh 24 each. But diamond appears at LL, making this combination impossible. You, therefore, force diamond (11).

5. Knowing diamond forces triangle (5) at LL and RL, circle (8) at RRL, and square (4) at RRR.

6. Knowing triangle forces ellipse (6) at RL.

Values

◆ = 11   △ = 5   ⊘ = 8   ▦ = 4

◖ = 6

Key	Sample Sequence
R = right arm **L** = left arm **C** = center arm	RRRL = (from the top) right arm, right arm of that arm, right arm of that arm, left arm of that arm.

Name _____

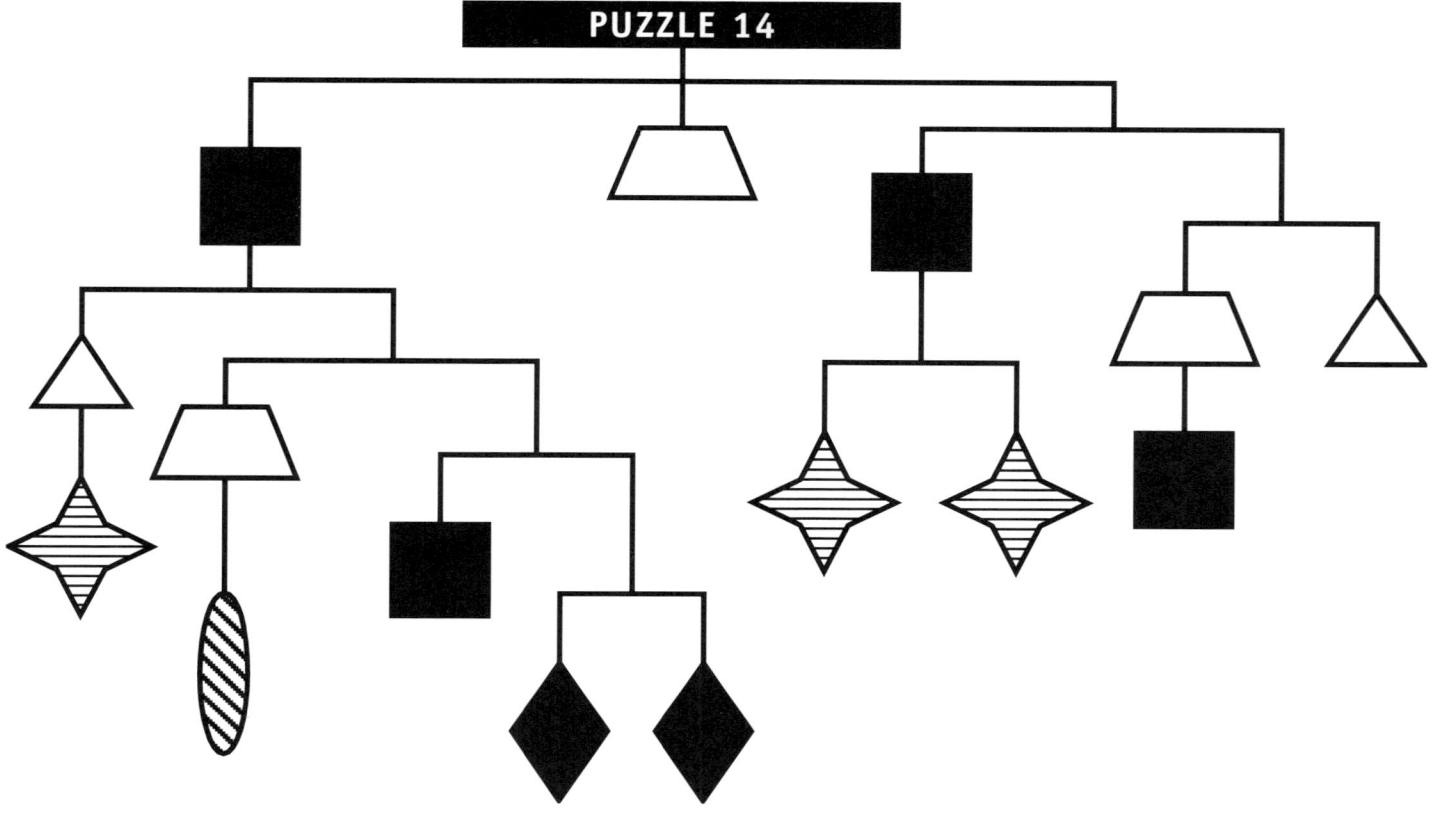

Discover the value of each of the shapes.
The total weight is 77.

IN THE BALANCE Grades 4–6

Solution

1. Trapezoid must be odd, because when subtracted from 77, it yields an even number.

2. Using an If-Then table to list possible trapezoid:right arm (R) combinations, you get 1:38, 5:36, 9:34, 13:32, 17:30. (You can eliminate all R possibilities that are odd, since R must be further divided.) But RR splits again, so only 5:18 and 13:16 are possible. But trapezoid cannot weigh 13 or greater at RRL because it would force RR to weigh too much. Therefore, trapezoid equals 5.

3. Knowing trapezoid, you can, by breaking down the R and L halves, learn triangle (9) at RRR.

4. Knowing the above forces square (4) at RRL and then star (7) at RLL.

5. Knowing the above forces diamond (2) at LRRR and ellipse (3) at LRL.

Values

$\square$ = 5 $\triangle$ = 9   ■ = 4   ◈ = 7

◆ = 2   ⬭ = 3

Key	Sample Sequence
R = right arm **L** = left arm **C** = center arm	RRRL = (from the top) right arm, right arm of that arm, right arm of that arm, left arm of that arm.

PUZZLE 15

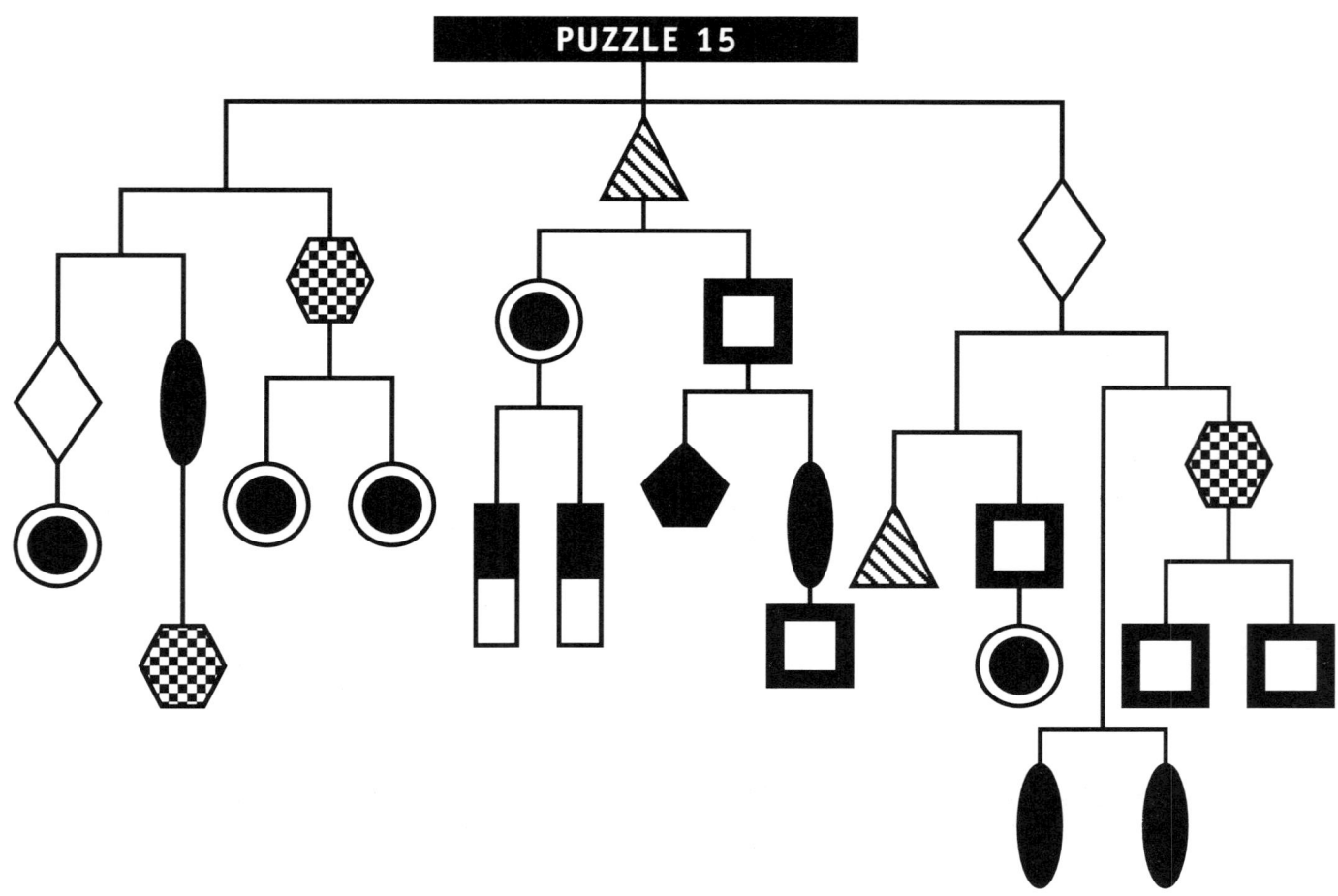

Discover the value of each of the shapes. The total weight is 180. Each of the three arms is equal in weight.

32

Solution

1. Triangle (C), diamond (R), and hexagon (LR) are even because, when subtracted from an even number (60), they each yield an even number.

2. Using an If-Then table to list possible hexagon:ellipse combinations (at LLR), you find 14:1, 12:3, 10:5, 8:7, 6:9, 4:11, and 2:13.

3. The two ellipses (at RRL) form half the weight of RR. Therefore, four ellipses would form the whole weight of RR. Since RR equals LR, you know that the total weight of R (60) equals diamond plus 8 ellipses.

4. Using an If-Then table to list possible hexagon:ellipse:diamond combinations then, you find 14:1:52, 12:3:36, 10:5:20, and 8:7:4. (Higher ellipse weights would be too heavy for R's weight of 60.) But the first three of these possibilities are too heavy for LLL. Therefore, you force hexagon (8), ellipse (7), and diamond (4).

5. Knowing diamond forces triangle (14) at RLL.

6. Knowing hexagon forces circle (11) at LRL and LRR.

7. Knowing triangle and circle forces rectangle (6) at CLL and CLR and square (3) at RLR.

8. Knowing ellipse and square forces pentagon (10) at CRL.

Values

 = 8   ⬭ = 7   ◇ = 4   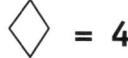 = 14

◉ = 11   ▮ = 6   ◻ = 3   ⬟ = 10

Key	Sample Sequence
R = right arm **L** = left arm **C** = center arm	RRRL = (from the top) right arm, right arm of that arm, right arm of that arm, left arm of that arm.

PUZZLE 16

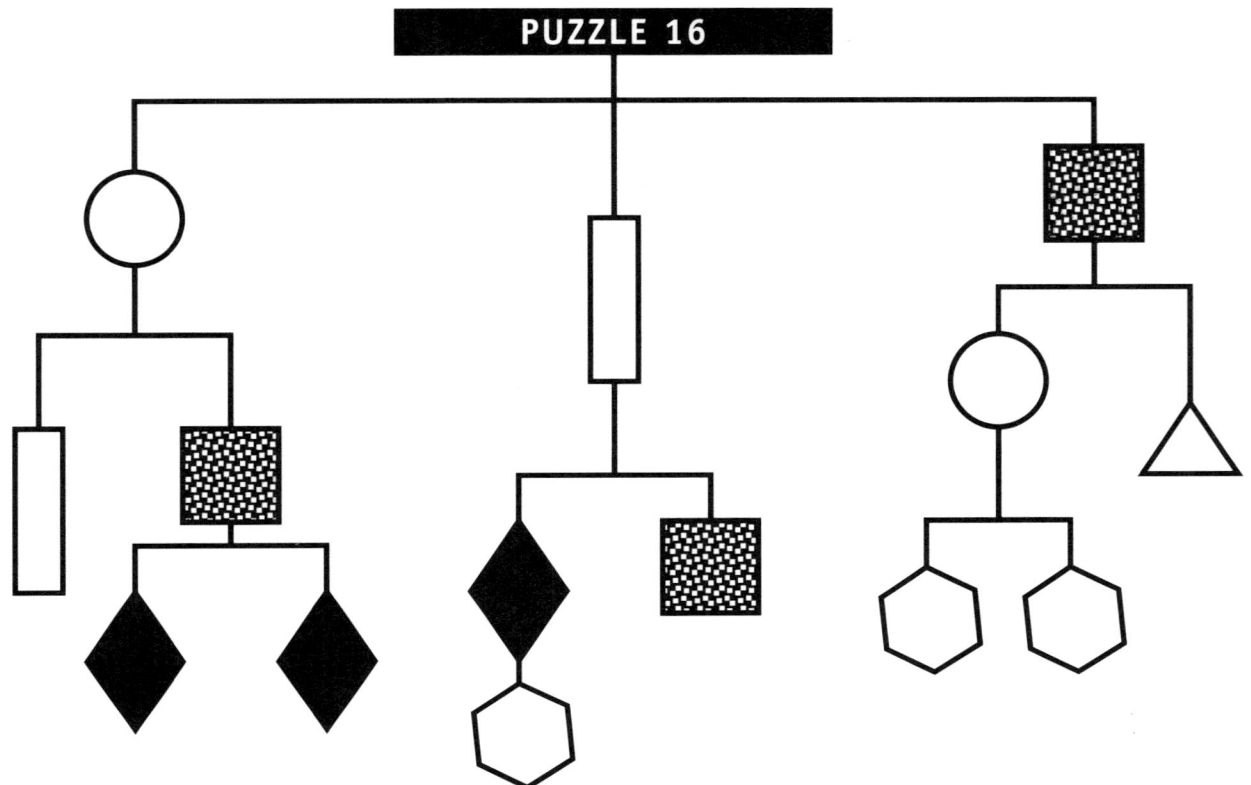

Discover the value of each of the shapes. The total weight is 57. Each of the three arms is equal in weight. Additional clue:

⬜ **is a multiple of three.**

IN THE BALANCE   Grades 4–6

Solution

1. By dividing the puzzle weight by three, you learn that each arm weighs 19.

2. Circle, rectangle, and square are odd because when subtracted from an odd number (19), they each yield a number that can be halved.

3. Rectangle is an odd (step 2) multiple of three (second clue). That means the rectangle can be 3 or 9. (If it were 15, it would weigh too much in LL.) Using an If-Then table to list possible rectangle:square combinations in C, you find 3:8 and 9:5. But you know that square is odd, eliminating 3:8 and forcing rectangle (9) and square (5).

4. Knowing rectangle and square forces diamond (2) at LR.

5. Knowing the above forces circle (1) at L.

6. Knowing diamond forces hexagon (3) at CL.

7. Knowing the above forces triangle (7) at RR.

Values

▯ = 9   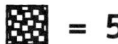 = 5   ◆ = 2   ◯ = 1

⬡ = 3   △ = 7

Key	Sample Sequence
R = right arm **L** = left arm **C** = center arm	RRRL = (from the top) right arm, right arm of that arm, right arm of that arm, left arm of that arm.

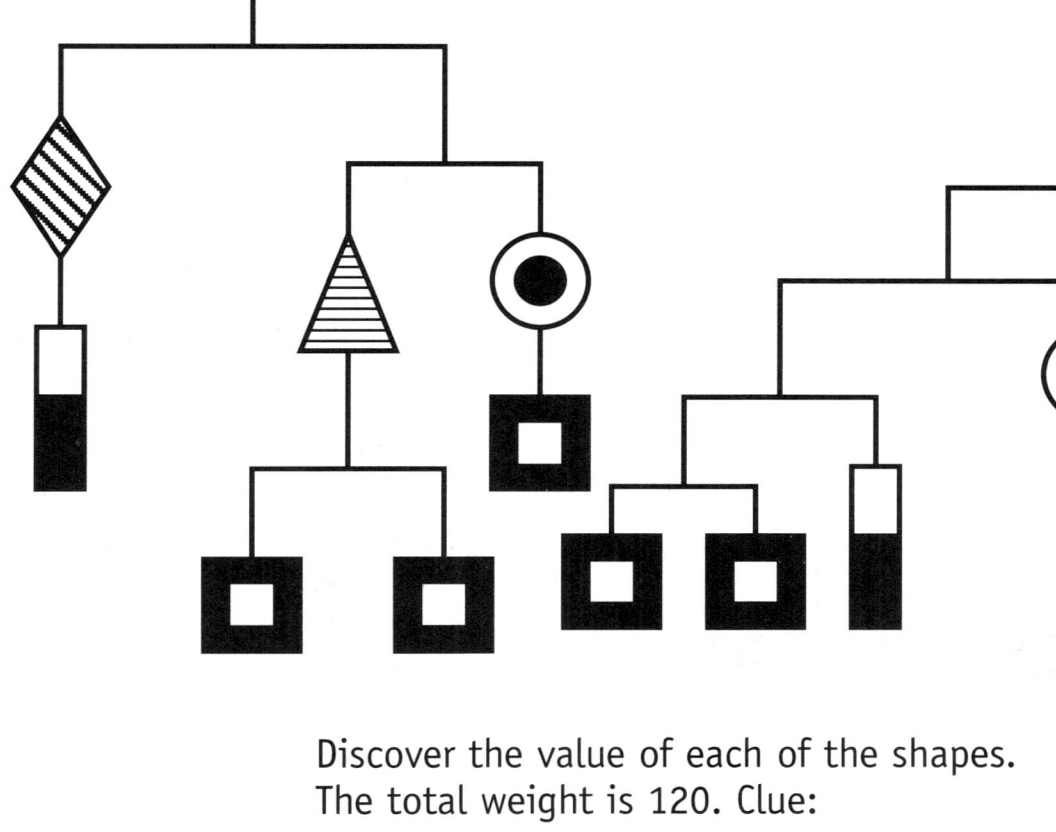

PUZZLE 17

Discover the value of each of the shapes.
The total weight is 120. Clue:

All shapes are multiples of ■.

Solution

1. Circle is even because, when subtracted from 60 at R, it yields an even number. Triangle (at LRL) is odd because when subtracted from 15, it yields a number that can be halved.

2. Using an If-Then table to list possible circle:square combinations that total 15 (at LRR), you find 2:13, 4:11, 6:9, 8:7, 10:5, 12:3, and 14:1.

3. But square, at RLLLL and RLLLR, cannot weigh 5 or greater because then the overall weight of R would exceed 60. Therefore, you force square (3), rectangle (6), circle (12), and diamond (24).

4. Knowing the above forces triangle (9) at LRL.

Values

□ = 3   ▮ = 6   ◉ = 12   ◈ = 24

△ = 9

Key	Sample Sequence
R = right arm **L** = left arm **C** = center arm	RRRL = (from the top) right arm, right arm of that arm, right arm of that arm, left arm of that arm.

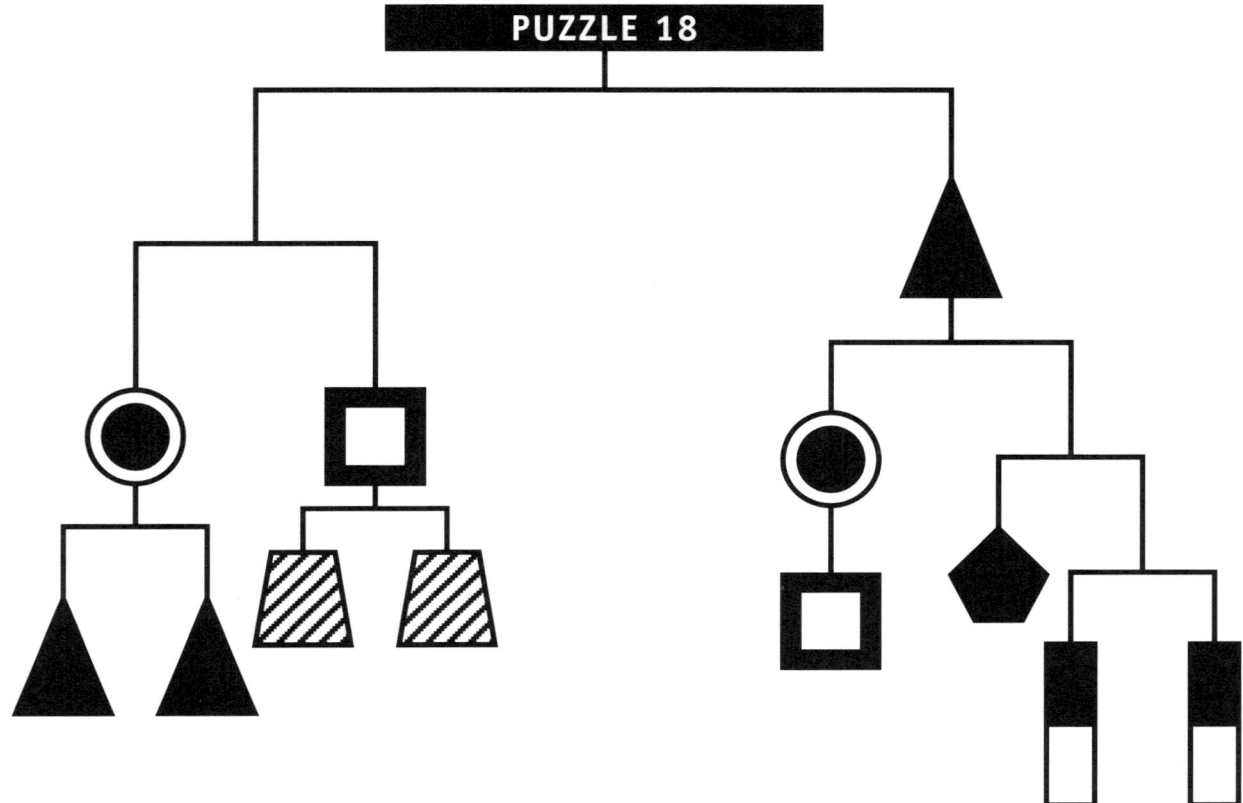

PUZZLE 18

Discover the value of each of the shapes. The total weight is 84. There are three even and three odd weights.

Permission is given by the publisher to the purchasing teacher or parent to reproduce this page for classroom or home use only.

IN THE BALANCE Grades 4–6

©Creative Publications® **38**

Solution

1. Triangle at R must be even because, when subtracted from 42, it yields an even difference.

2. Using an If-Then table to list triangle:circle combinations that total 21 at LL, you find 2:17, 4:13, 6:9, 8:5, and 10:1.

3. Using an If-Then table to list triangle:pentagon:rectangle combinations (that work at R and RR), you find 2:10:5 and 10:8:4.

4. Using circle possibilities from above, list possible circle:square:trapezoid combinations (that work at RL and LR), and you find 17:3:9 and 1:15:3.

5. Given the clue, and using all the above information, you learn circle (1), triangle (10), pentagon (8), rectangle (4), square (15), and trapezoid (3).

Values

⊙ = 1   ▲ = 10   ⬟ = 8   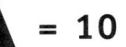 = 4

■ = 15    = 3

Key	Sample Sequence
R = right arm **L** = left arm **C** = center arm	RRRL = (from the top) right arm, right arm of that arm, right arm of that arm, left arm of that arm.

Name

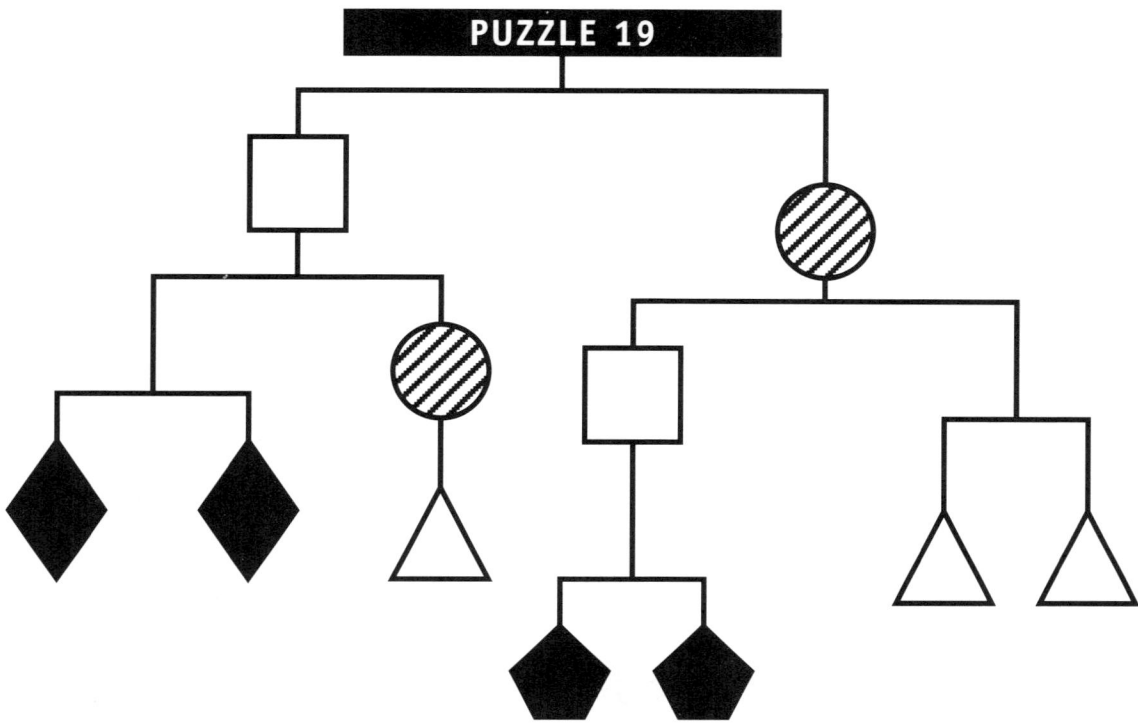

Discover the value of each of the shapes.
The total weight is 68. Clue:

$$\square + \blacklozenge = \oslash$$

IN THE BALANCE Grades 4–6

40

Solution

1. By halving the puzzle weight you learn each main arm weighs 34. Both the square and circle must be even because, when subtracted from 34, they yield an even number (a number that can be divided in half).

2. Square is limited, then, to 2, 4, 6, 8, 10, 12, 14, 16. (A value of 18 or higher would be too much for R, causing the arm to be more than 34.)

3. Using an If-Then table to list possible square:diamond combinations, you get only four "workable" possibilities: 2:8, 6:7, 10:6, 14:5. Square values that are multiples of 4 would yield fractional weights. (For example, a square weighing 4 would force a diamond weight of 7.5.)

4. Given the clue and knowing that the circle has an even-numbered value, you can eliminate the square:diamond combinations of 6:7 and 14:5 because they would yield odd-numbered values for the circle. You also can eliminate the 10:6 combination because the circle's value of 16, when added to that of the triangle in LR, would be too heavy to balance the two diamonds on LL. That leaves you with only one possibility: square (2) and diamond (8).

5. Knowing square and diamond, and given the clue, you learn circle (10) and force triangle (6).

6. Knowing square, triangle, and circle forces pentagon (5) in RL.

Values

□ = 2   ◆ = 8   ⊘ = 10   △ = 6

⬟ = 5

Key	Sample Sequence
R = right arm **L** = left arm **C** = center arm	RRRL = (from the top) right arm, right arm of that arm, right arm of that arm, left arm of that arm.

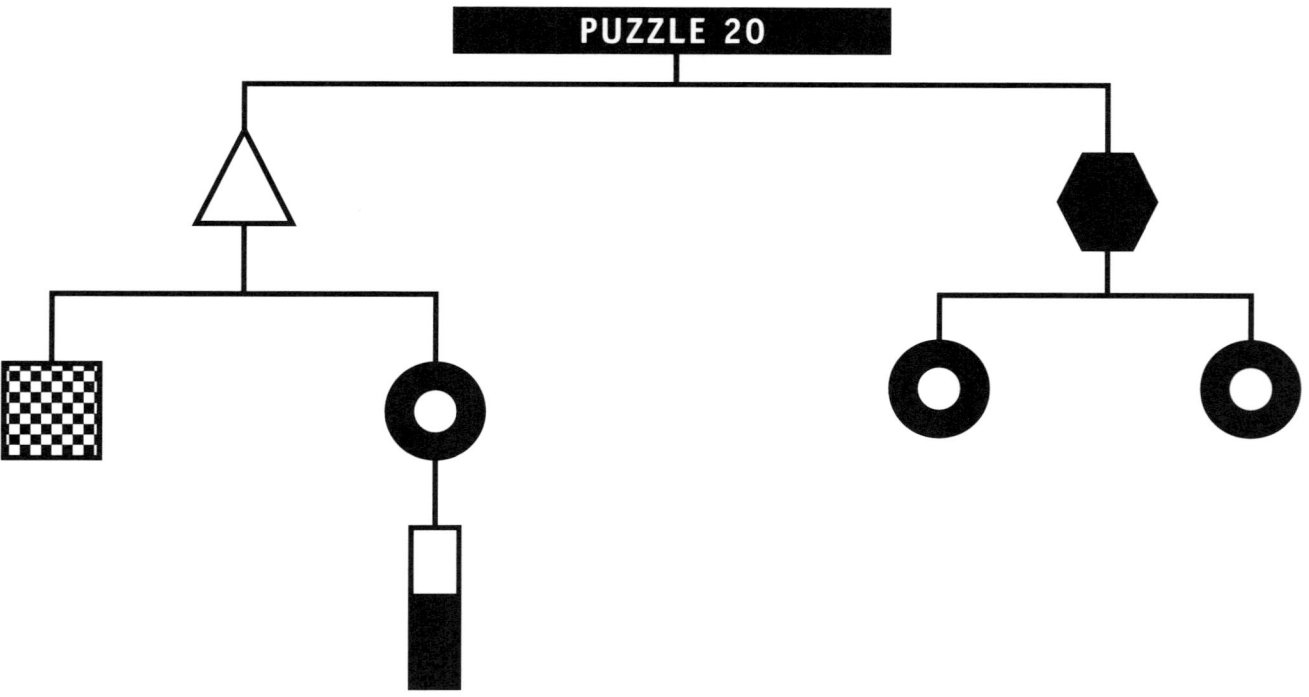

Discover the value of each of the shapes.
The total weight is 30. Clues:

$$3 \boxplus < \hexagon \qquad \triangle - \boxplus = \blacksquare$$

Solution

1. Triangle and hexagon must be odd because, when subtracted from an odd number (15), they each yield an even number.

2. Using an If-Then table to list possible triangle:square combinations (at L:LL), you find 1:7, 3:6, 5:5, 7:4, 9:3, 11:2, and 13:1. You eliminate 5:5 because no shapes can have the same value. Using the first clue, you can eliminate possible square values of 6 and 7, because they would make L too heavy. The second clue eliminates 11:2 and 13:1 because the difference would be too great for L. So you're left with possible triangle:square combinations of 7:4 and 9:3.

3. Considering then only square values of 4 or 3, you know the sum of circle plus rectangle (at LR) must be 4 or 3 respectively to balance.

4. Using the second clue again with the two triangle:square possibilities (step 2), you know that rectangle would equal 3 or 6. But a rectangle value of 6 would be too heavy (step 3). Therefore, you force rectangle (3), square (4), triangle (7).

5. Knowing the above forces circle (1) and hexagon (13).

Values

$\blacksquare$ = 3 $\boxtimes$ = 4 $\triangle$ = 7 $\bullet$ = 1

 = 13

Key	Sample Sequence
R = right arm **L** = left arm **C** = center arm	RRRL = (from the top) right arm, right arm of that arm, right arm of that arm, left arm of that arm.

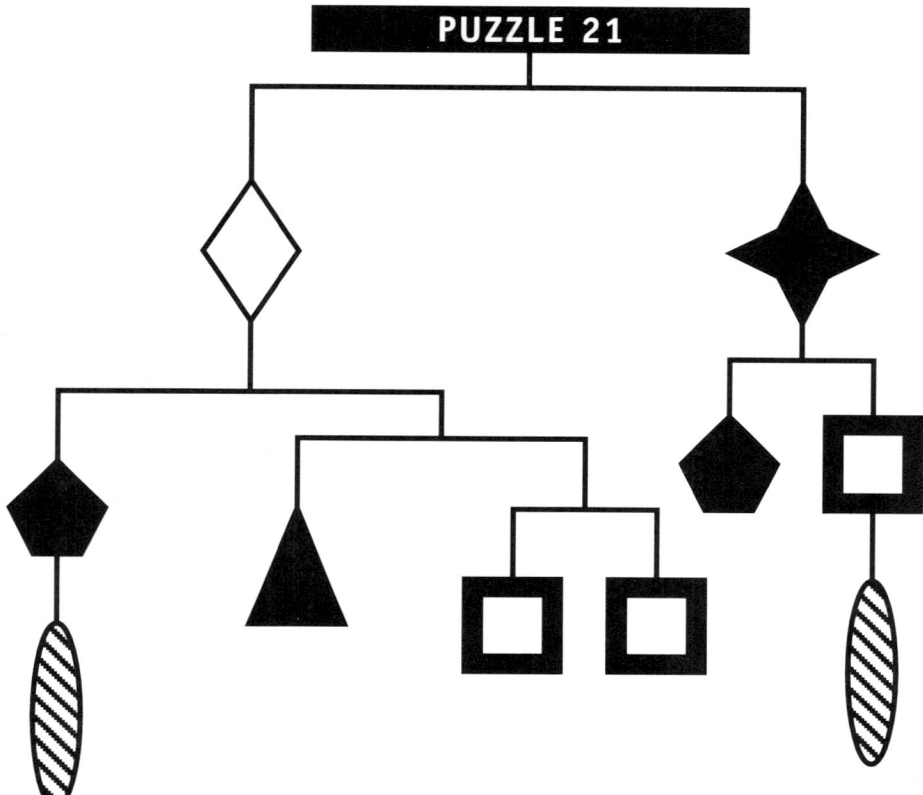

PUZZLE 21

Discover the value of each of the shapes.
The total weight is 62. Clue:

◇ − ■ is a prime number.

44

Solution

1. Diamond (at L) and star (at R) must be odd because, when subtracted from an odd number (31), they each yield even numbers.

2. Using an If-Then table to list possible square:triangle:diamond combinations (on L), you find 1:2:23, 2:4:15, and 3:6:7. A value higher than 3 at square would yield a weight too heavy for L.

3. Only the second of these possibilities fits the clue given, so you learn square (2), triangle (4), and diamond (15).

4. Using an If-Then table to list possible pentagon:ellipse combinations that will total 8 at LL, you find 1:7, 3:5, 5:3, and 7:1. (The other possibilities require numbers already used above.)

5. You know (since LL balances LR) that square (which you know is 2) plus ellipse equals pentagon. Checking back to the possible pentagon:ellipse combinations (step 4), you find that only 5:3 works. Therefore, you learn pentagon (5) and ellipse (3).

6. Knowing the above forces star (21) at R.

Values

■ = 2   ▲ = 4   ◇ = 15   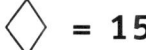 = 5

 = 3   ✦ = 21

Key	Sample Sequence
R = right arm **L** = left arm **C** = center arm	RRRL = (from the top) right arm, right arm of that arm, right arm of that arm, left arm of that arm.

PUZZLE 22

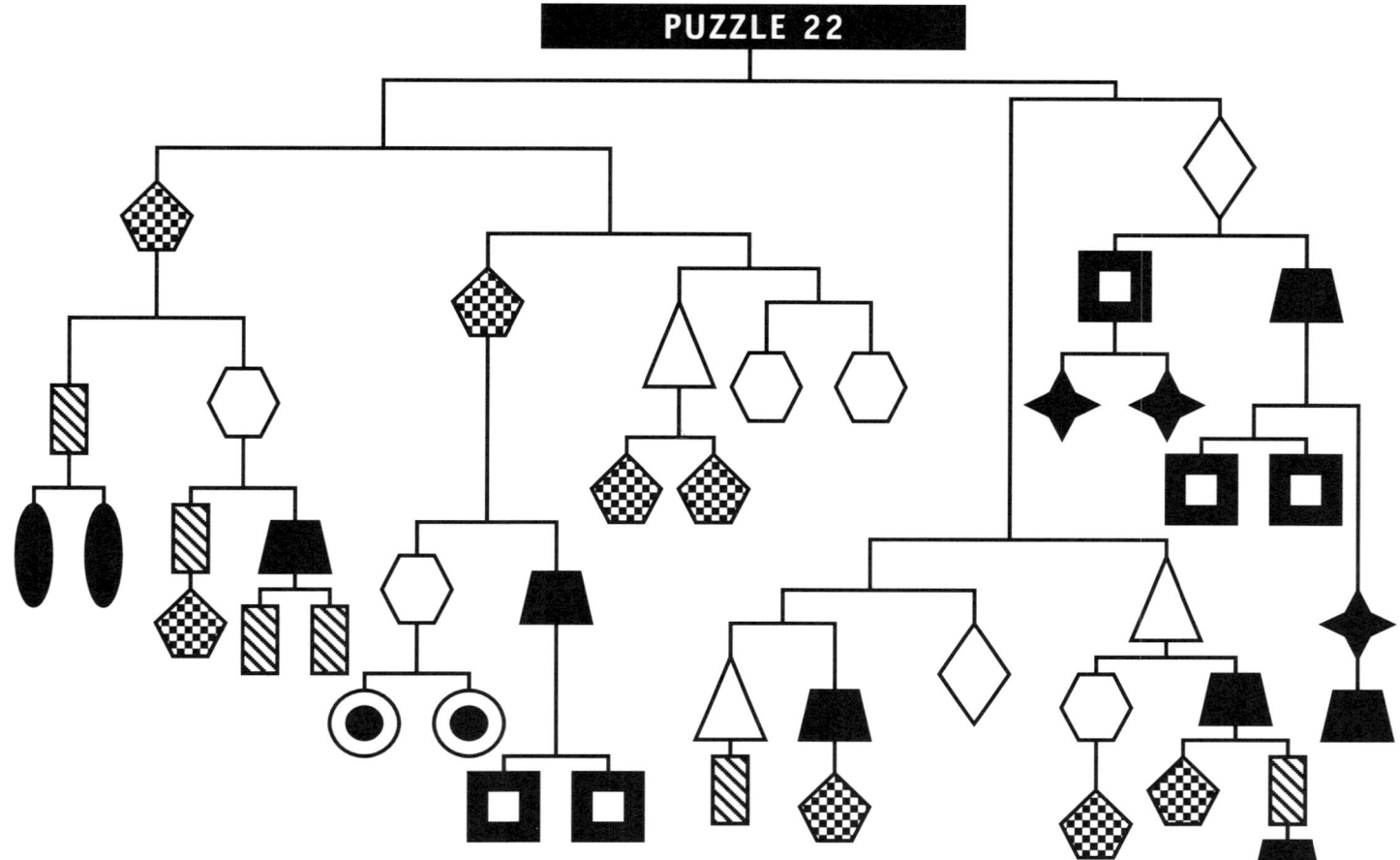

Discover the value of each of the shapes.
The total weight is 416. Clue:

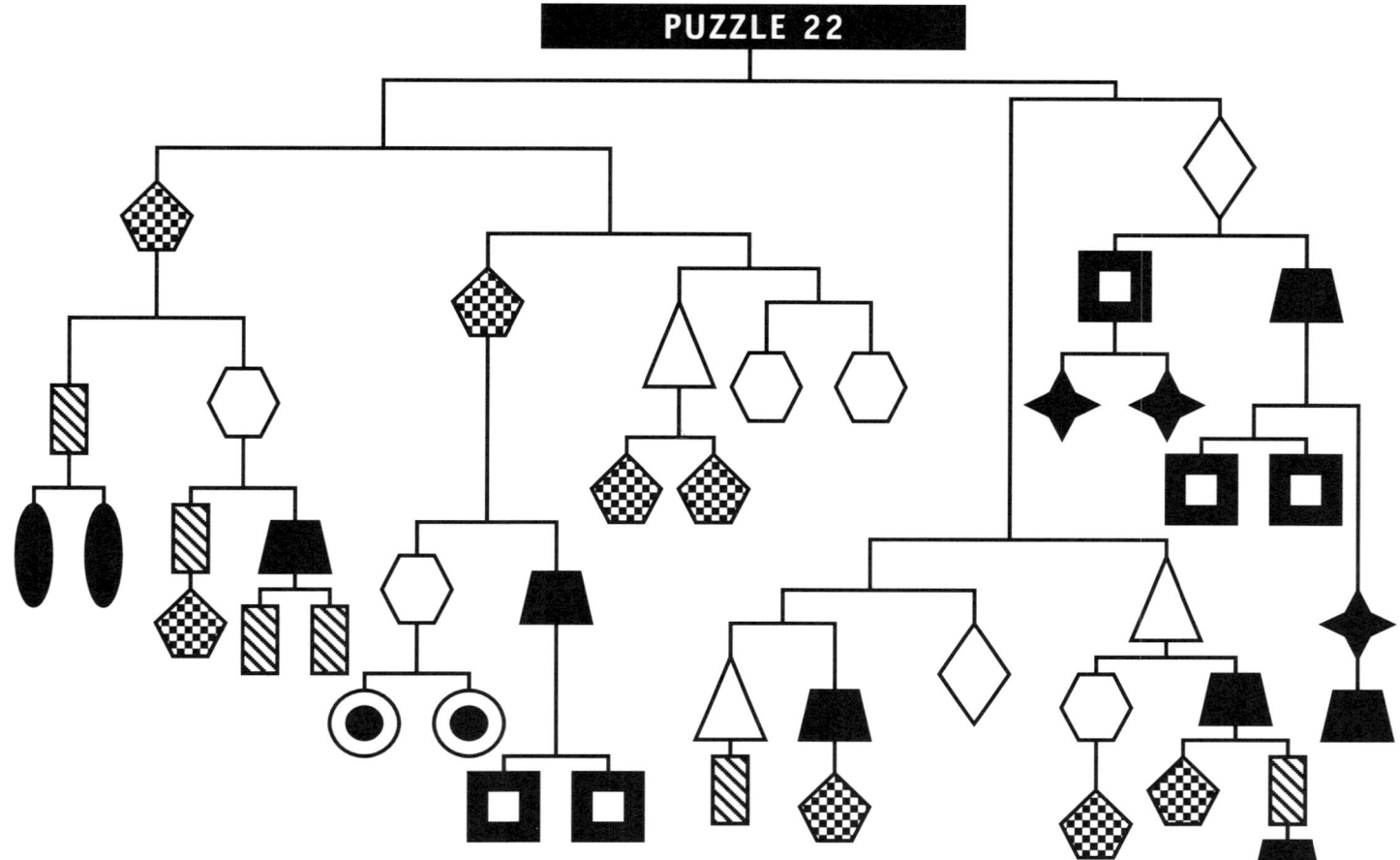 – △ = 1

IN THE BALANCE Grades 4–6

46

Solution

1. By halving the arms, you learn the value of the diamond (26) in RLLR and the hexagons (13 each) in LRRR.

2. On RLLLL, you know that the triangle and the rectangle weigh 13. Given the clue, you learn rectangle (7) and triangle (6).

3. Knowing triangle forces pentagon (10) on LRRL.

4. Knowing pentagon forces trapezoid (3) on RLRR.

5. Knowing rectangle, trapezoid, pentagon, and hexagon forces ellipse (20) in LLL.

6. Knowing pentagon and hexagon forces circle (4) in LRLL.

7. Knowing trapezoid forces square (9) in LRLR.

8. Knowing square forces the star (15) in RRRR.

Values

◇ = 26   ⬡ = 13   ◩ = 7   △ = 6

♦ = 10   ▲ = 3   ⬭ = 20   ◉ = 4

▢ = 9   ✦ = 15

Key	Sample Sequence
R = right arm **L** = left arm **C** = center arm	RRRL = (from the top) right arm, right arm of that arm, right arm of that arm, left arm of that arm.

PUZZLE 23

Discover the value of each of the shapes.
The total weight is 44. Clue:

◖ > ◇

48

Solution

1. Triangle (LL) and circle (RL) must be odd because, when subtracted from an odd number (11), they each yield a number that can be halved.

2. Using an If-Then table to list possible circle:triangle combinations at RL, you find that there are three: 1:5, 5:3, and 9:1.

3. Using the triangle possibilities from step 2, now use an If-Then table to list triangle:hexagon possibilities at LL and LLR. Only one possibility works: 3:2. Therefore, triangle equals 3 and hexagon equals 2.

4. Knowing triangle (3) forces circle (5) at RL.

5. Knowing hexagon (2) at LLR forces rectangle (4) at LLL.

6. Knowing hexagon (2) at RR forces trapezoid (9).

7. Since you've already assigned 2, 3, 4, 5, and 9, the only diamond:ellipse possibilities at LR would be 1:10 and 10:1. The clue forces ellipse (10) and diamond (1).

Values

▲ = 3   ✿ = 2   ◉ = 5   ▮ = 4

▱ = 9   ⬮ = 10   ◇ = 1

Key	Sample Sequence
R = right arm **L** = left arm **C** = center arm	RRRL = (from the top) right arm, right arm of that arm, right arm of that arm, left arm of that arm.

PUZZLE 24

Discover the value of each of the shapes. The total weight is 51. Each of the three arms is equal in weight. Additional clues:

 have consecutively numbered values.   ◇ > △

Solution

1. Breaking down the puzzle weight into thirds, you learn the weight of each arm (17).

2. Diamond and hexagon must be odd because, when subtracted from an odd number (17), they each yield an even number.

3. Using an If-Then table to list possible diamond:circle combinations, you find 1:8, 3:7, 5:6, 7:5, 9:4, and 11:3. Only two of these possibilities fit the second clue given: 5:6 and 7:5.

4. Using an If-Then table to list possible square:star combinations for a circle value of 6 (step 3), you find a possible 4:2 that fits the second clue. Circle value 5 yields no solutions that fit the second clue given. Therefore, you learn square (4), star (2), diamond (5), and circle (6).

5. Knowing star and using the third clue forces hexagon (15) and triangle (1) at C.

6. Knowing circle and square forces ellipse (7) at R.

Values

▨ = 4   ◇ = 2   ◈ = 5   ○ = 6

⬢ = 15   △ = 1   ⬮ = 7

Key	Sample Sequence
R = right arm **L** = left arm **C** = center arm	RRRL = (from the top) right arm, right arm of that arm, right arm of that arm, left arm of that arm.

PUZZLE 25

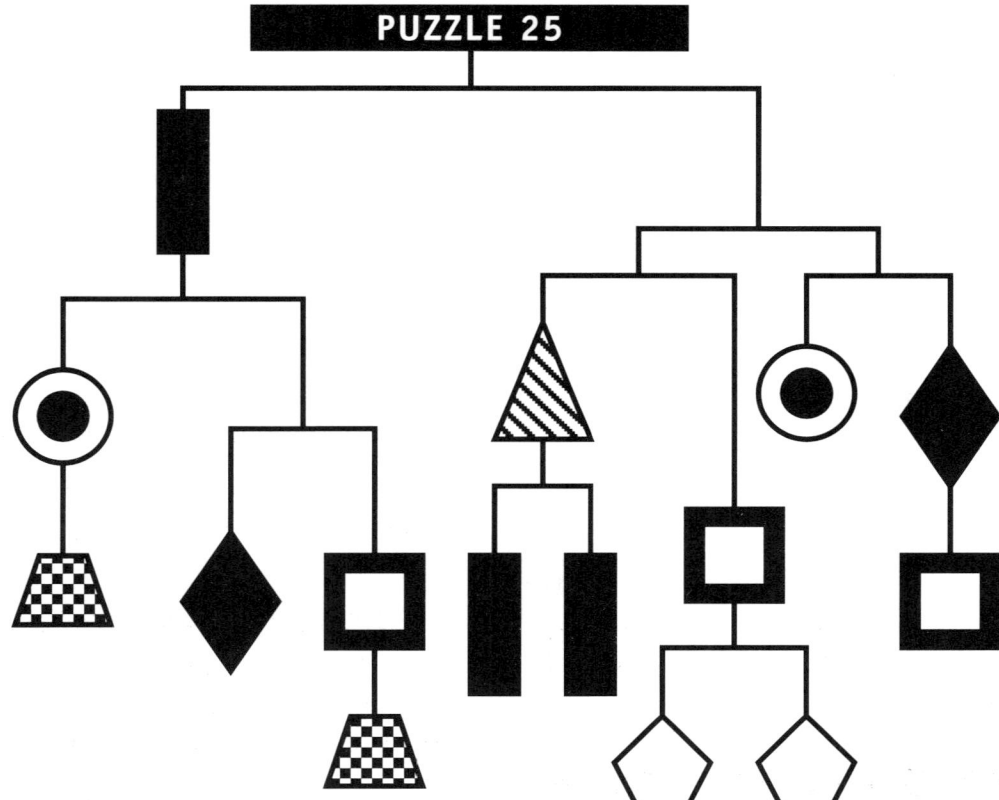

Discover the value of each of the shapes.
The total weight is 88. Clue:

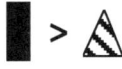

Solution

1. By halving the weights of the arms, you learn circle (11) at RRL.

2. Triangle (at RLL) and square (RLR) are odd because, when subtracted from an odd weight (11), they each yield an even weight.

3. Rectangle (at L) must be even because, when subtracted from an even weight, it yields an even weight.

4. Using an If-then table to list possible triangle:rectangle combinations at RL, you find 3:4 and 7:2. The clue forces the first combination and you learn triangle (3) and rectangle (4).

5. Knowing the above forces trapezoid (9) at LL.

6. Knowing the above forces diamond (10) at LRL and square (1) at LRR.

7. Knowing the above forces pentagon (5) at RLR.

Values

⊙ = 11   ◬ = 3   ▮ = 4   ▦ = 9

◆ = 10   ◨ = 1   ⬠ = 5

Key	Sample Sequence
R = right arm **L** = left arm **C** = center arm	RRRL = (from the top) right arm, right arm of that arm, right arm of that arm, left arm of that arm.

ROMPECABEZAS 1

Encuentra el valor de cada forma. El valor total
es 36. Todas las formas valen menos de 10.
Pista adicional:

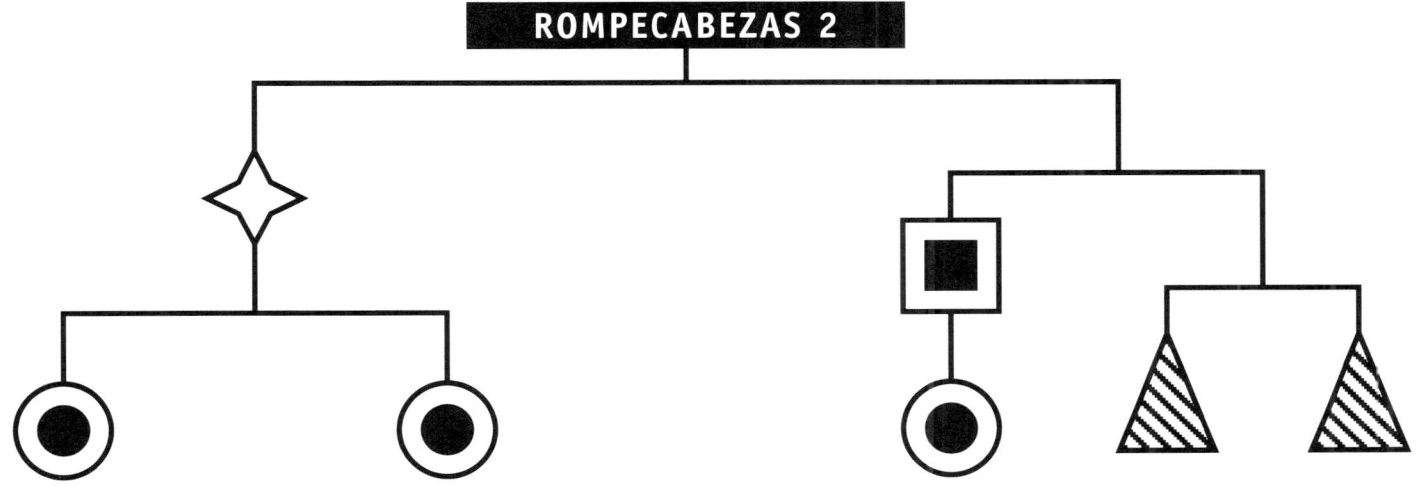

ROMPECABEZAS 2

Encuentra el valor de cada forma.
El valor total es 32. Pista:

$$\diamond - 2 = \blacksquare + \odot$$

55

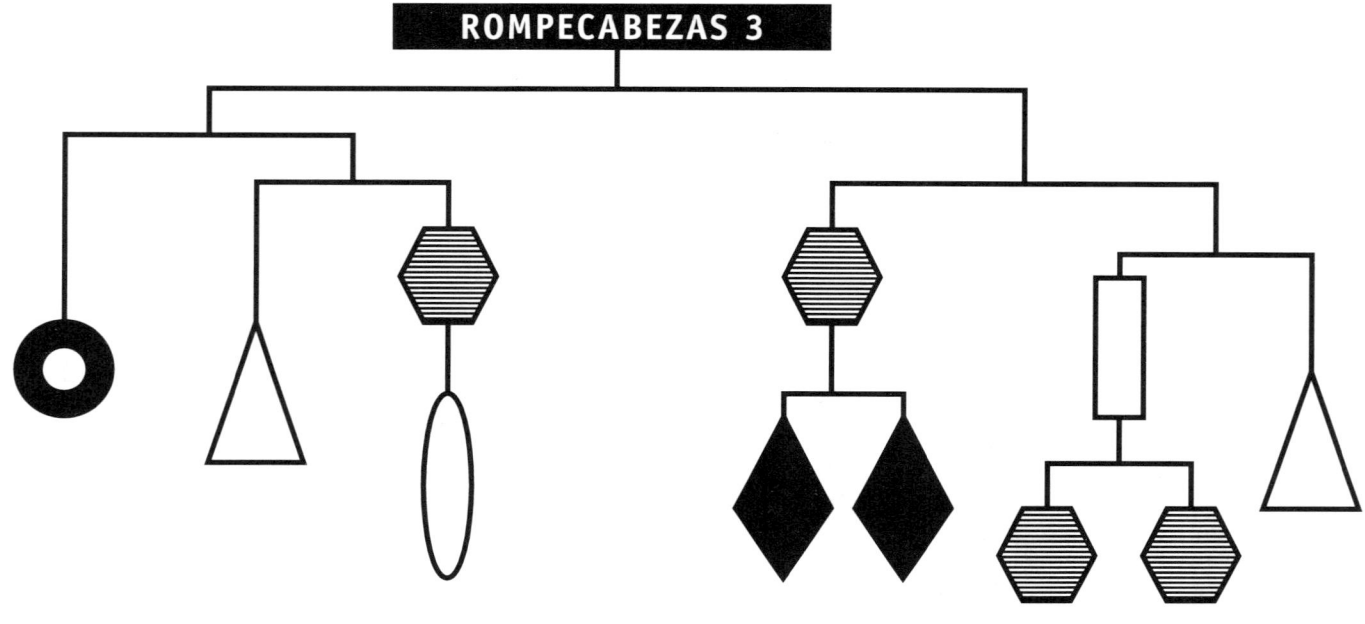

ROMPECABEZAS 3

Encuentra el valor de cada forma.
El valor total es 40.

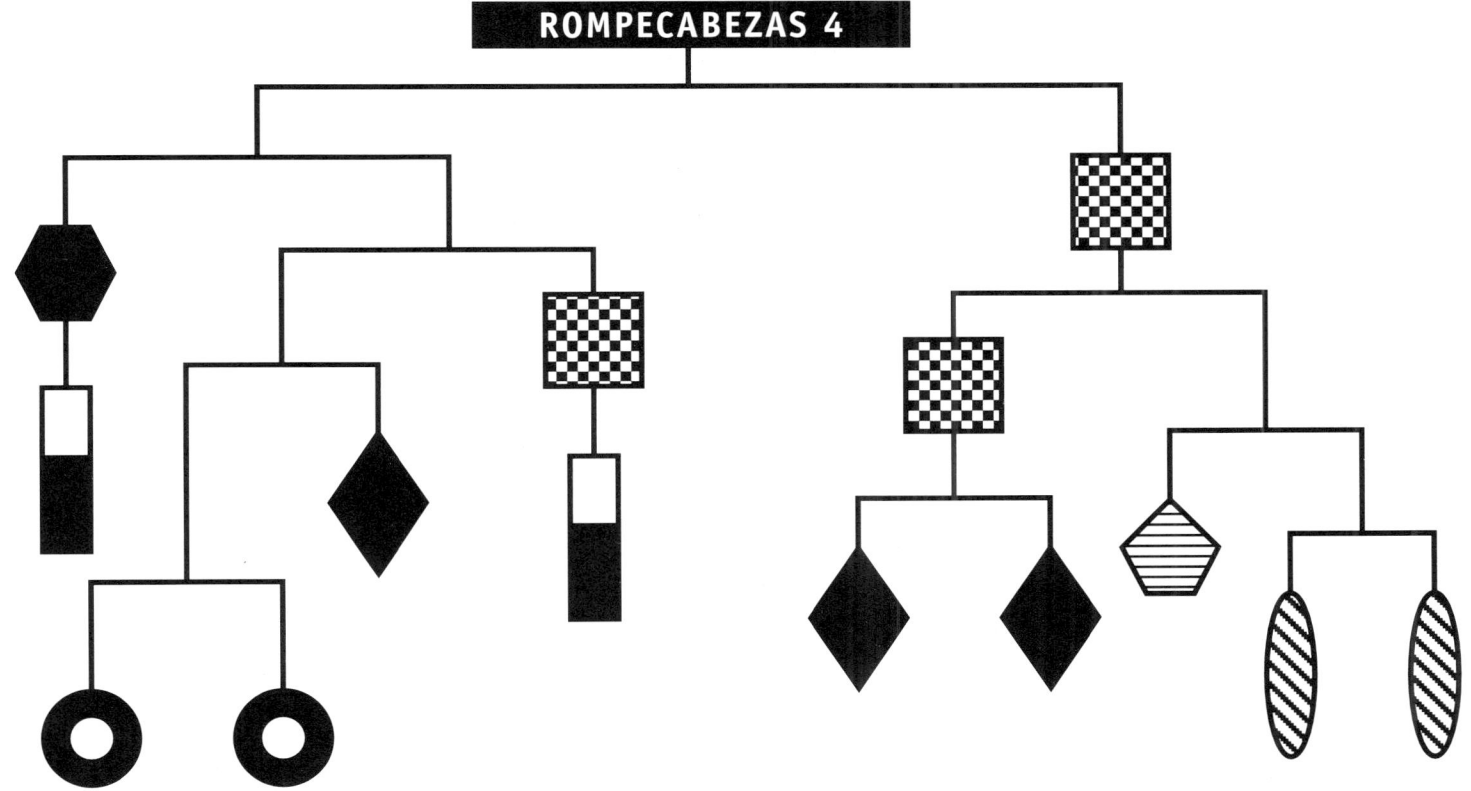

ROMPECABEZAS 4

Encuentra el valor de cada forma.
El valor total es 96.

IN THE BALANCE   Grades 4–6

©Creative Publications®   **57**

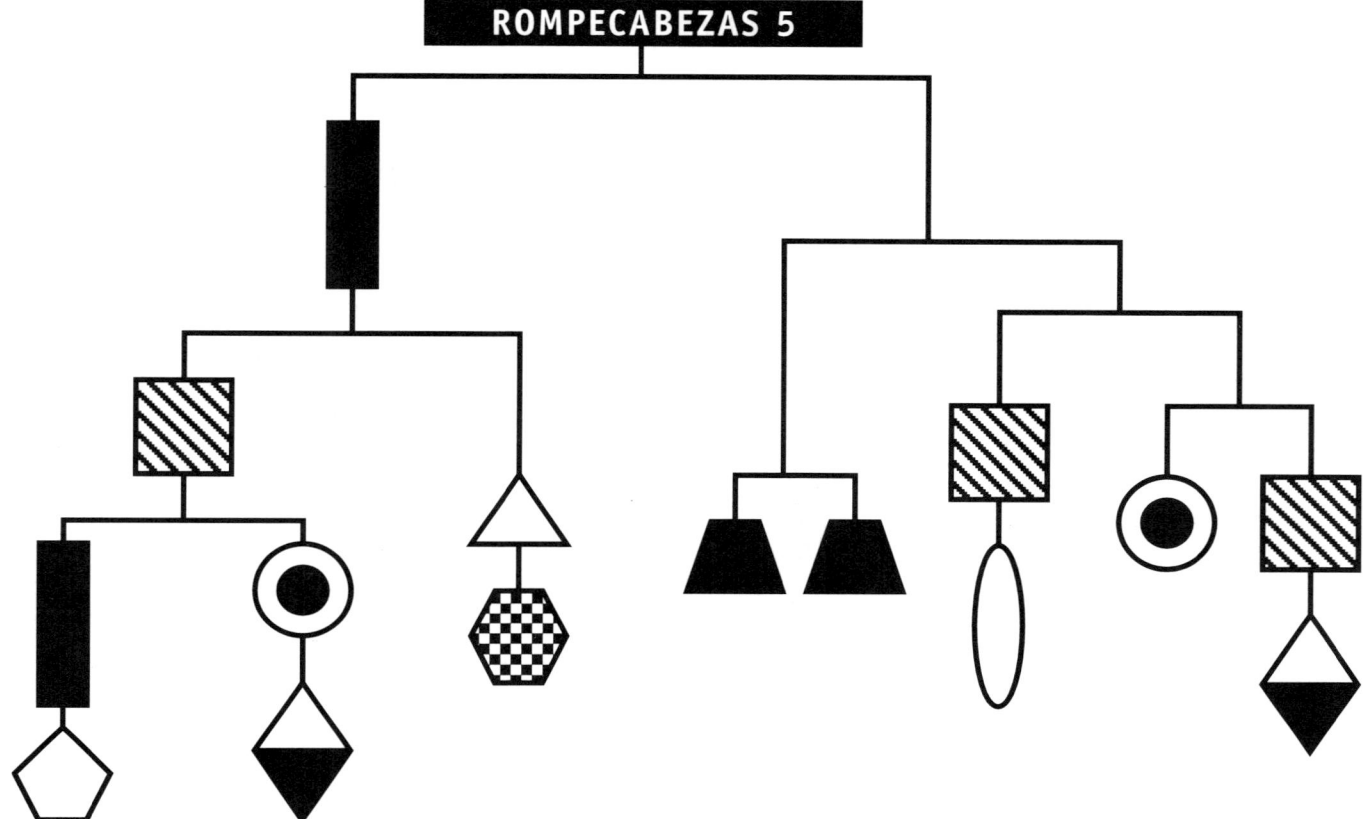

Encuentra el valor de cada forma. El valor total
es 80. Sólo una forma vale más de nueve.
Pistas adicionales:

◆ + 1 = ▨   △ < ⬢

IN THE BALANCE   Grades 4–6

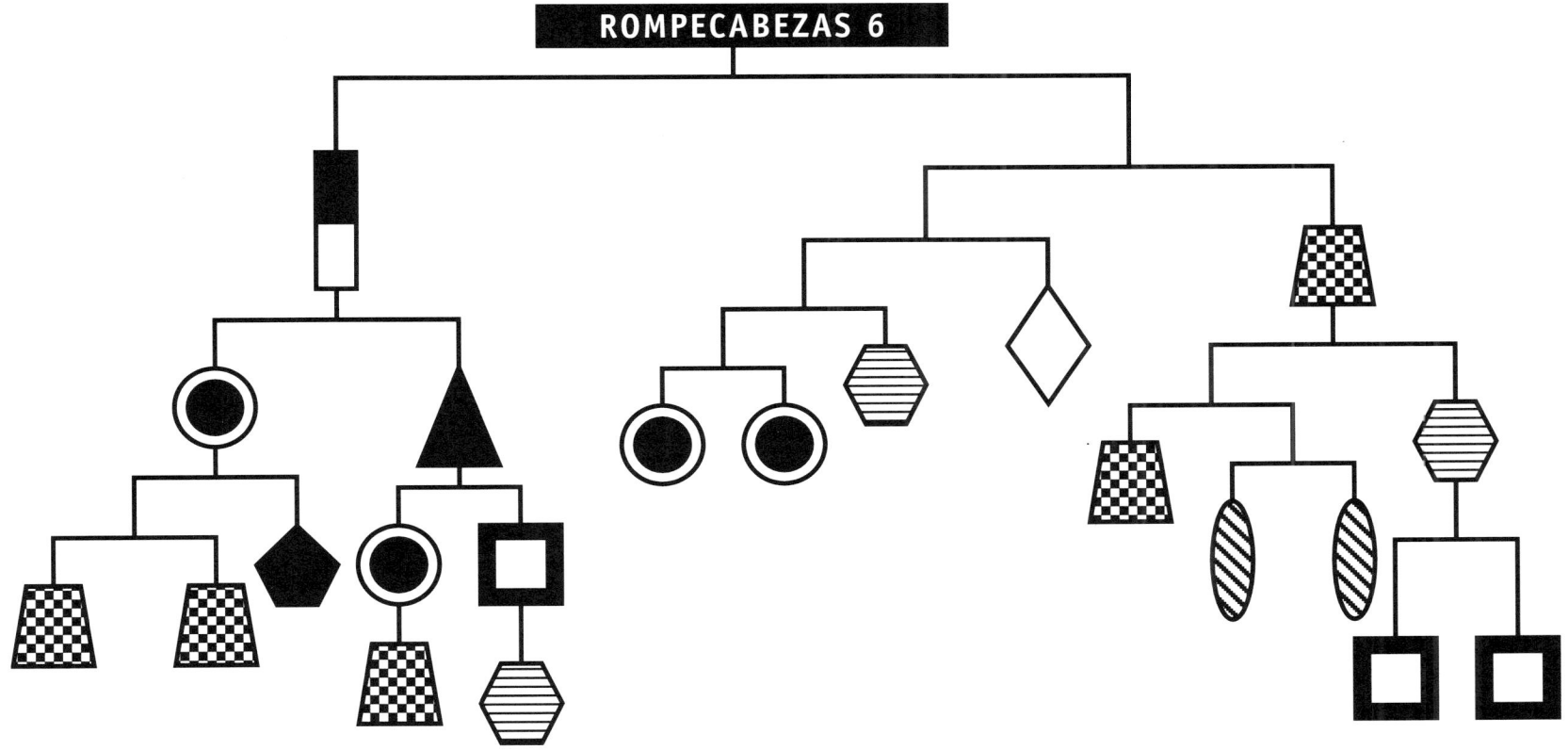

ROMPECABEZAS 6

Encuentra el valor de cada forma.
El valor total es 160.

IN THE BALANCE Grades 4–6

59

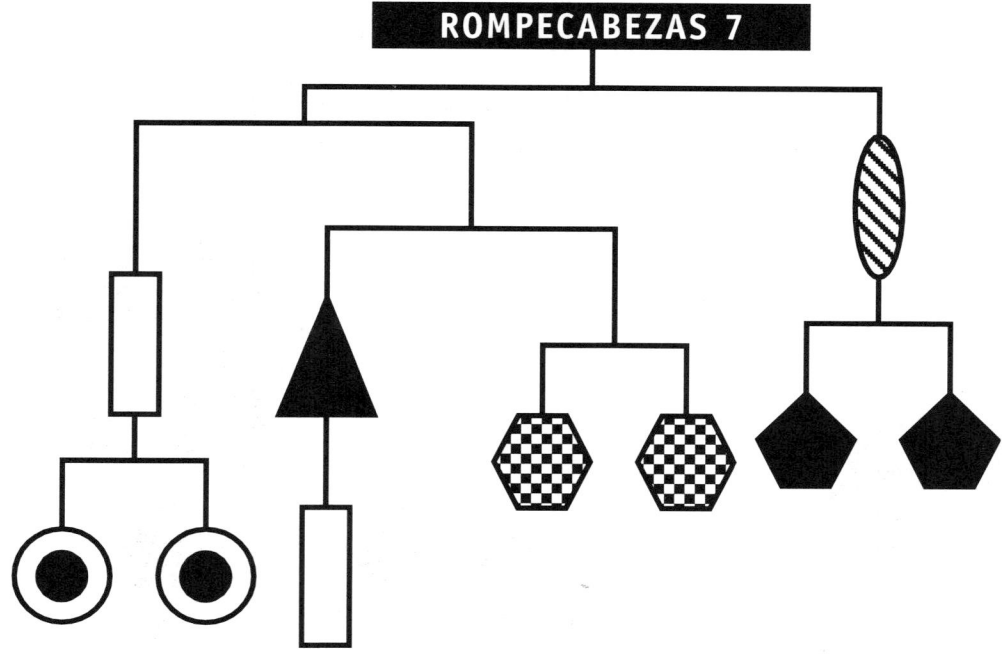

Encuentra el valor de cada forma.
El valor total es 48. Pistas:

Nombre ____________________

ROMPECABEZAS 8

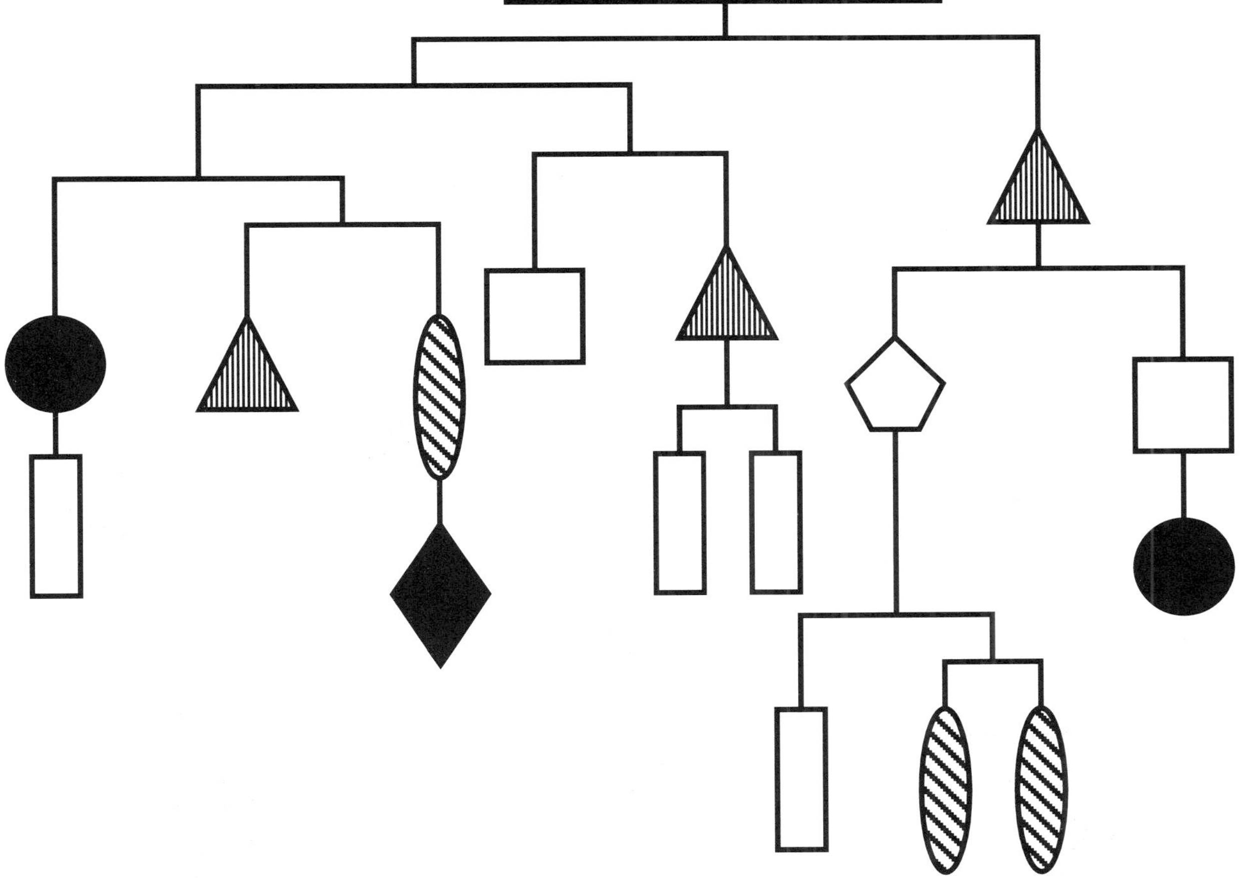

Encuentra el valor de cada forma.
El valor total es 64. Pista:

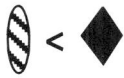

Nombre _____

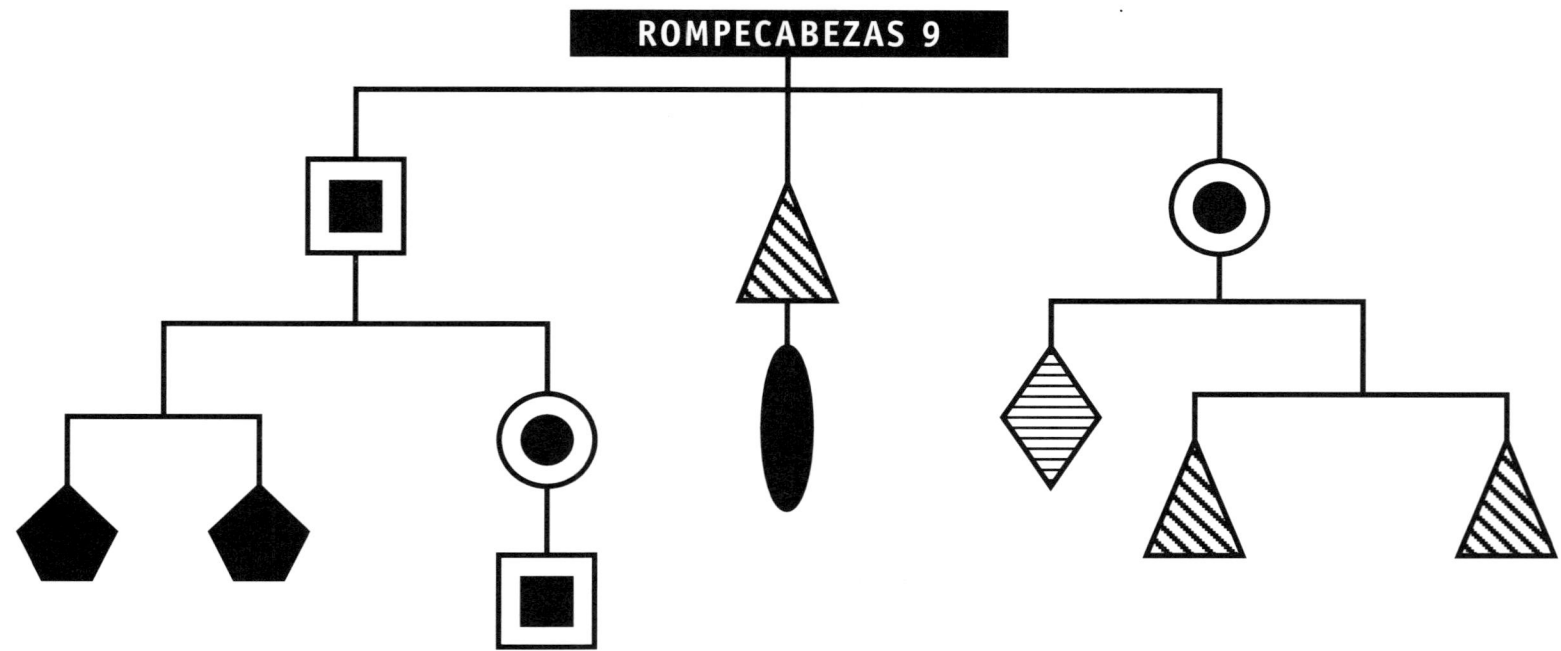

ROMPECABEZAS 9

Encuentra el valor de cada forma. El valor total
es 129. Cada rama tiene el mismo valor.
Pista adicional:

⬭ = 5 ◼

62

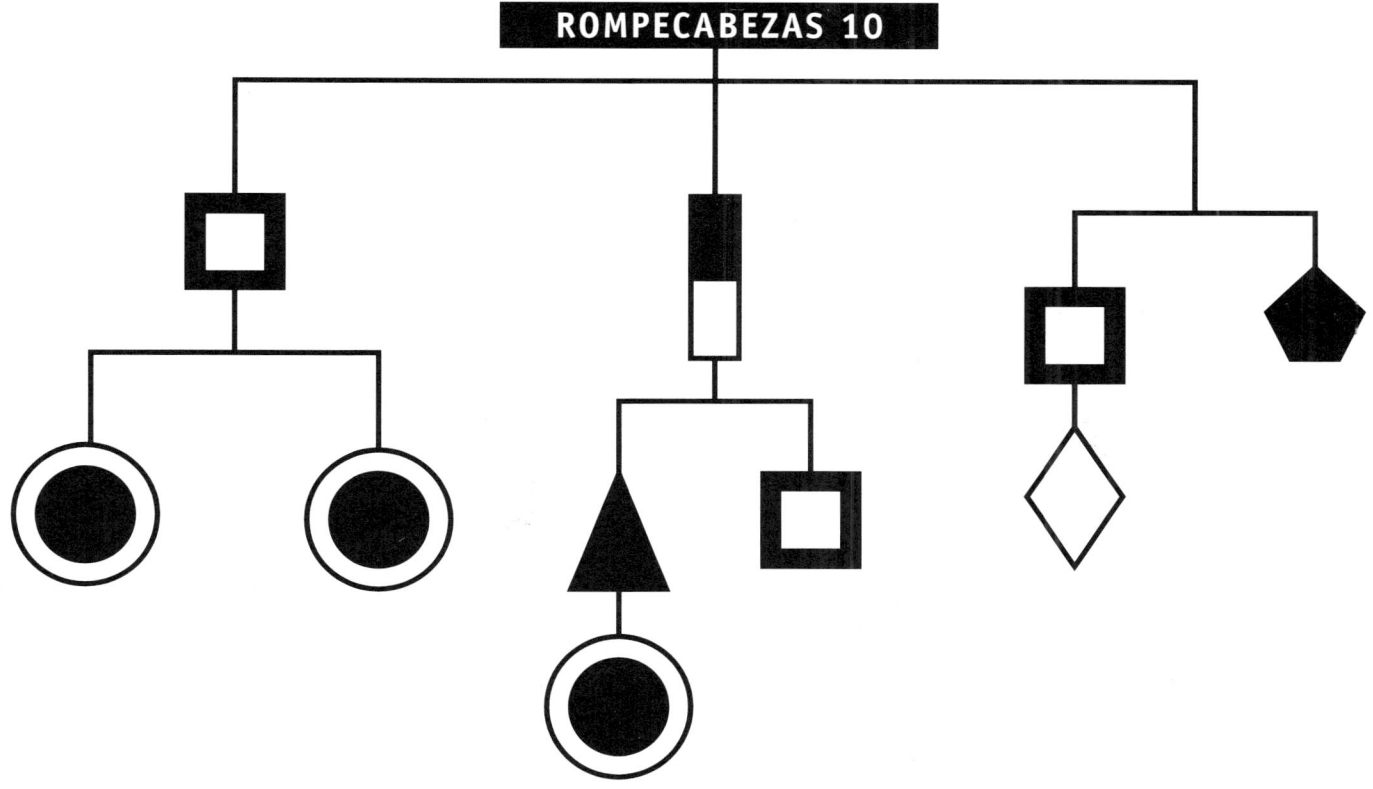

ROMPECABEZAS 10

Encuentra el valor de cada forma. El valor total
es 54. Cada rama tiene el mismo valor.

63

Nombre _____

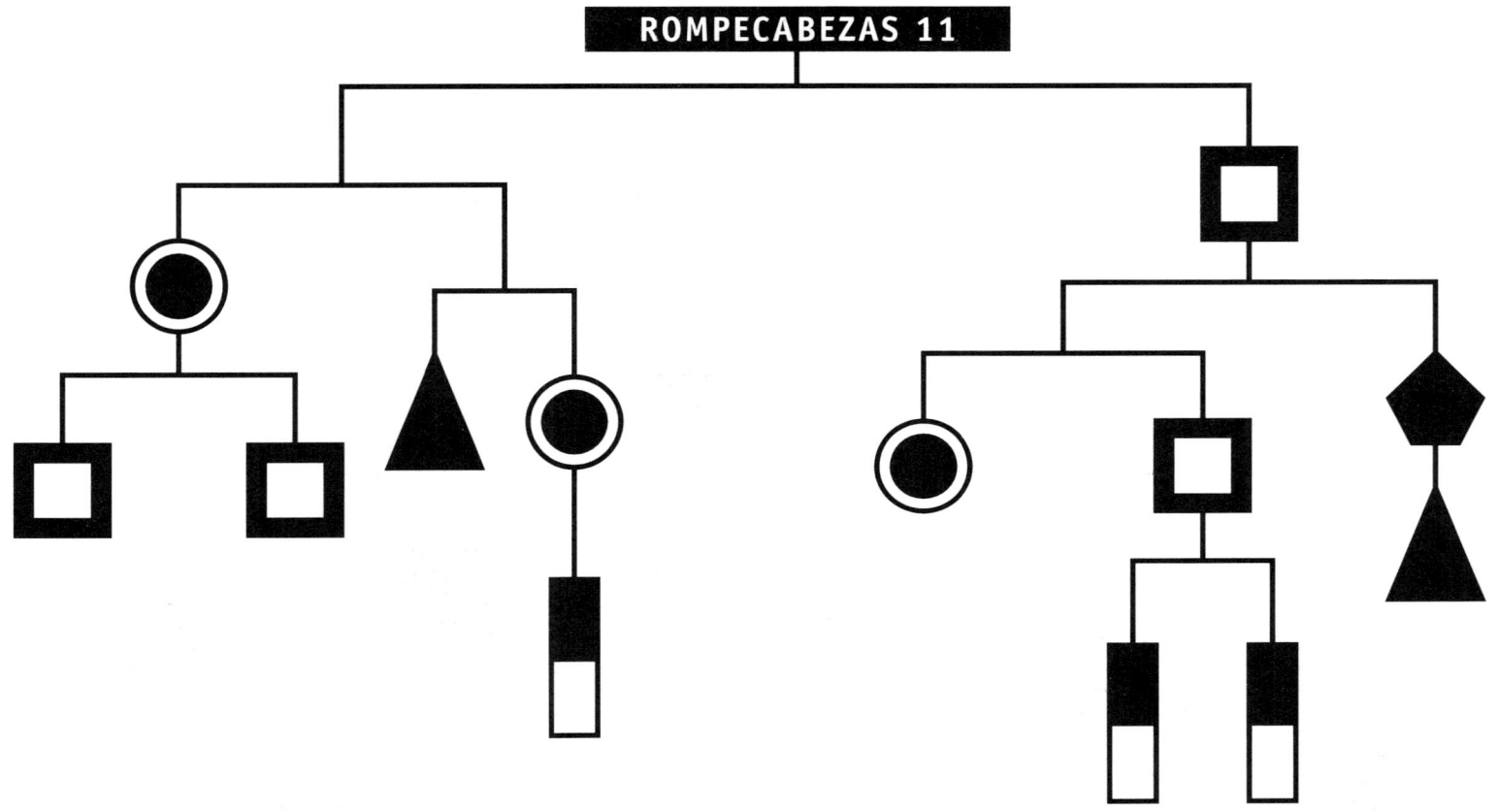

Encuentra el valor de cada forma.
El valor total es 56.

64

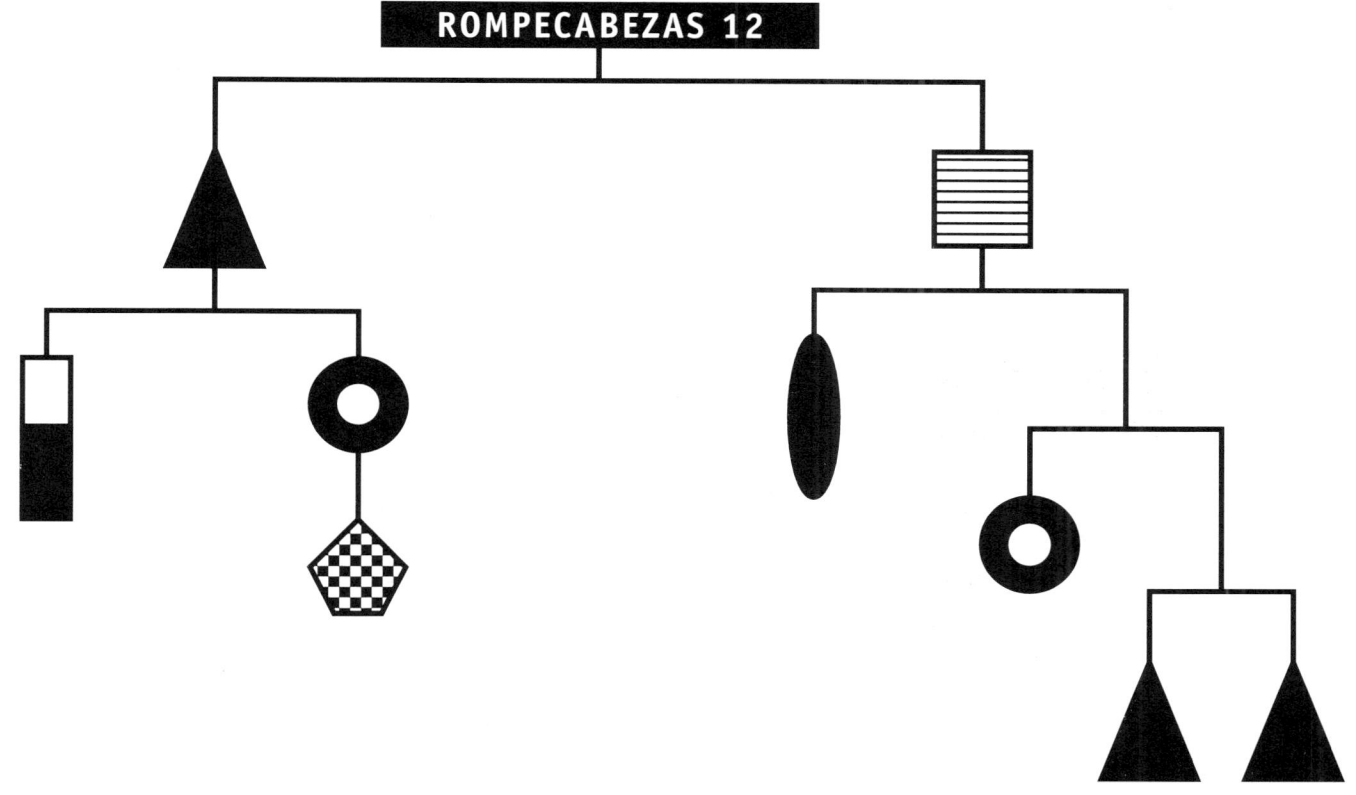

Encuentra el valor de cada forma.
El valor total es 50. Pista:

IN THE BALANCE Grades 4–6

65

Nombre _____

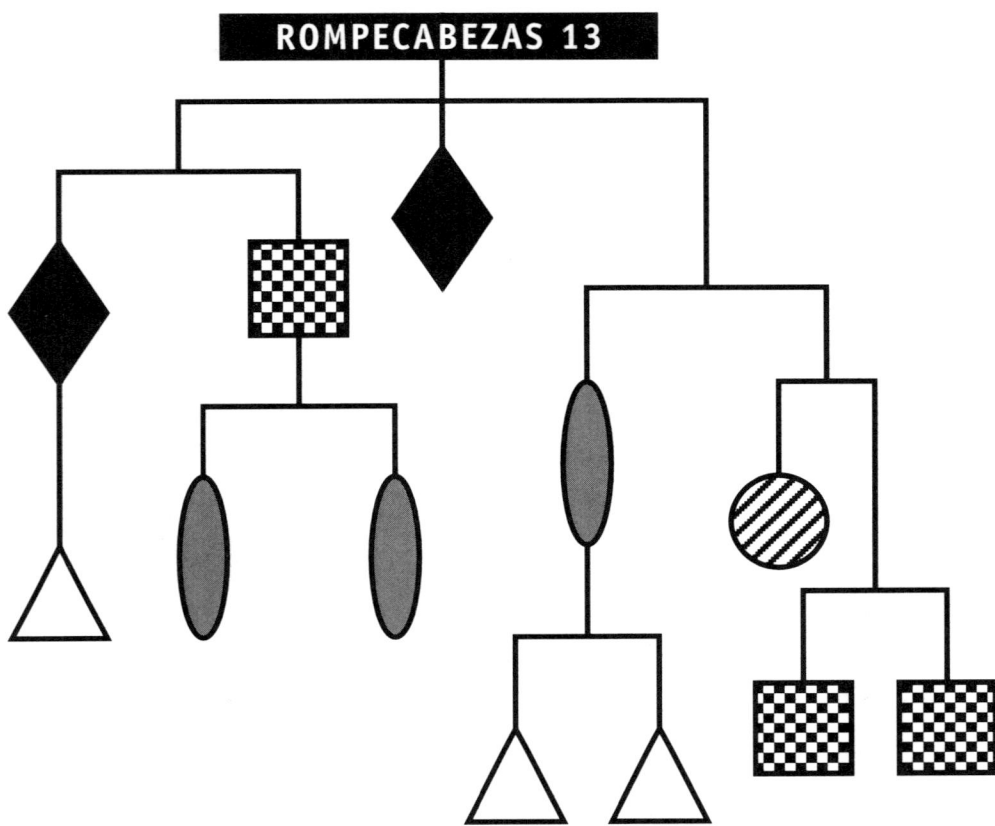

ROMPECABEZAS 13

Encuentra el valor de cada forma.
El valor total es 75.

 66

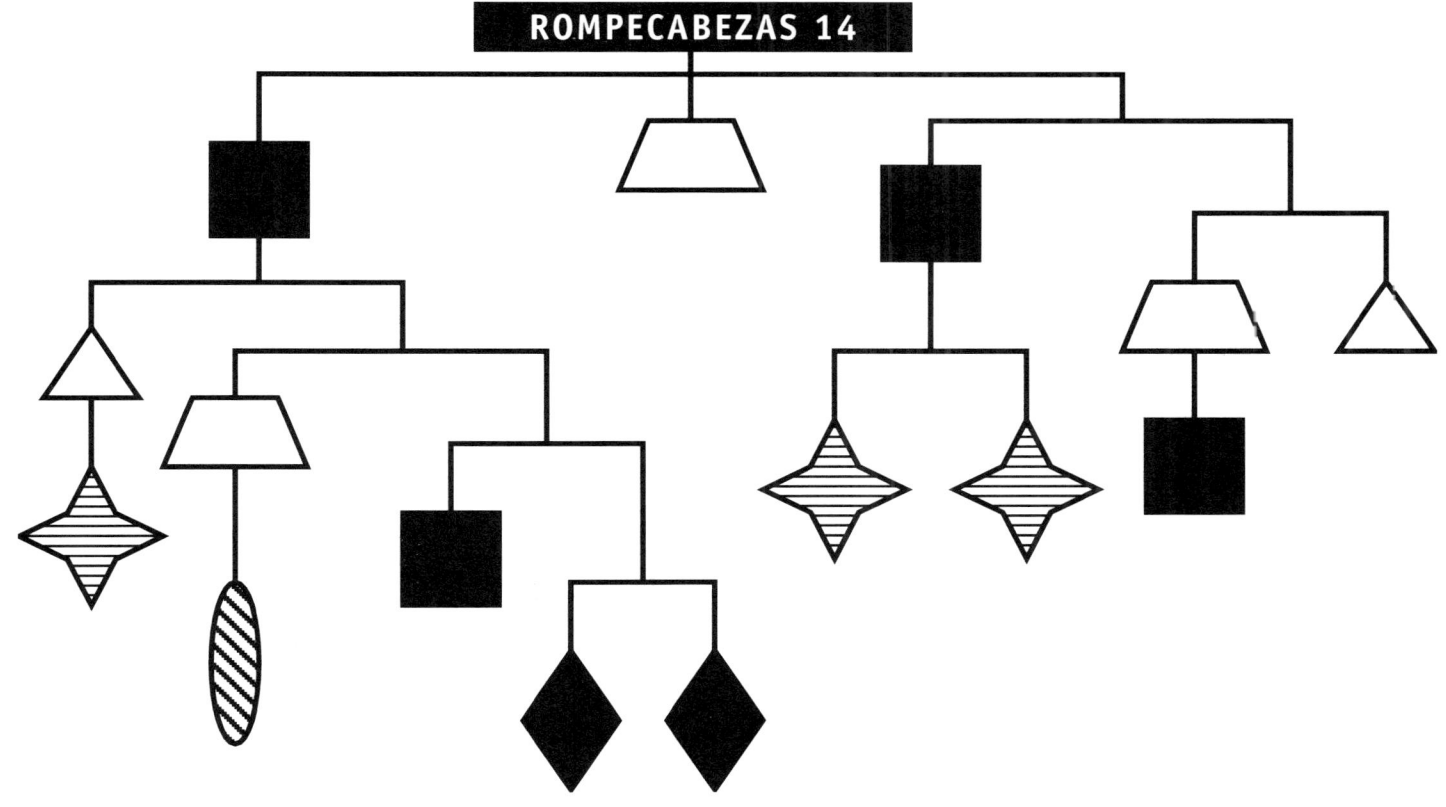

ROMPECABEZAS 14

Encuentra el valor de cada forma.
El valor total es 77.

Permission is given by the publisher to the purchasing teacher or
parent to reproduce this page for classroom or home use only.

IN THE BALANCE   Grades 4–6

© Creative Publications®   **67**

ROMPECABEZAS 15

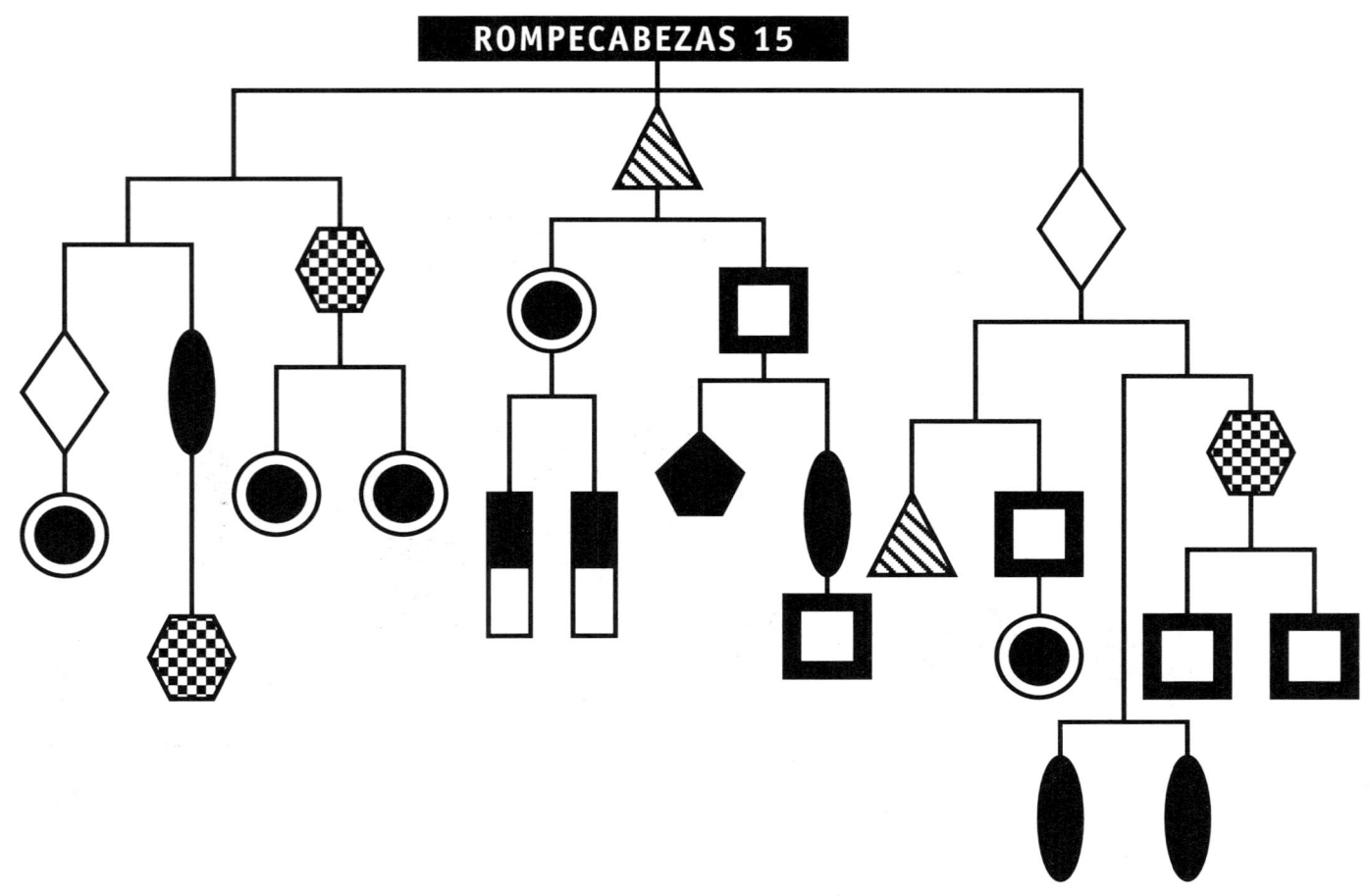

Encuentra el valor de cada forma. El valor total
es 180. Cada rama tiene el mismo valor.

68

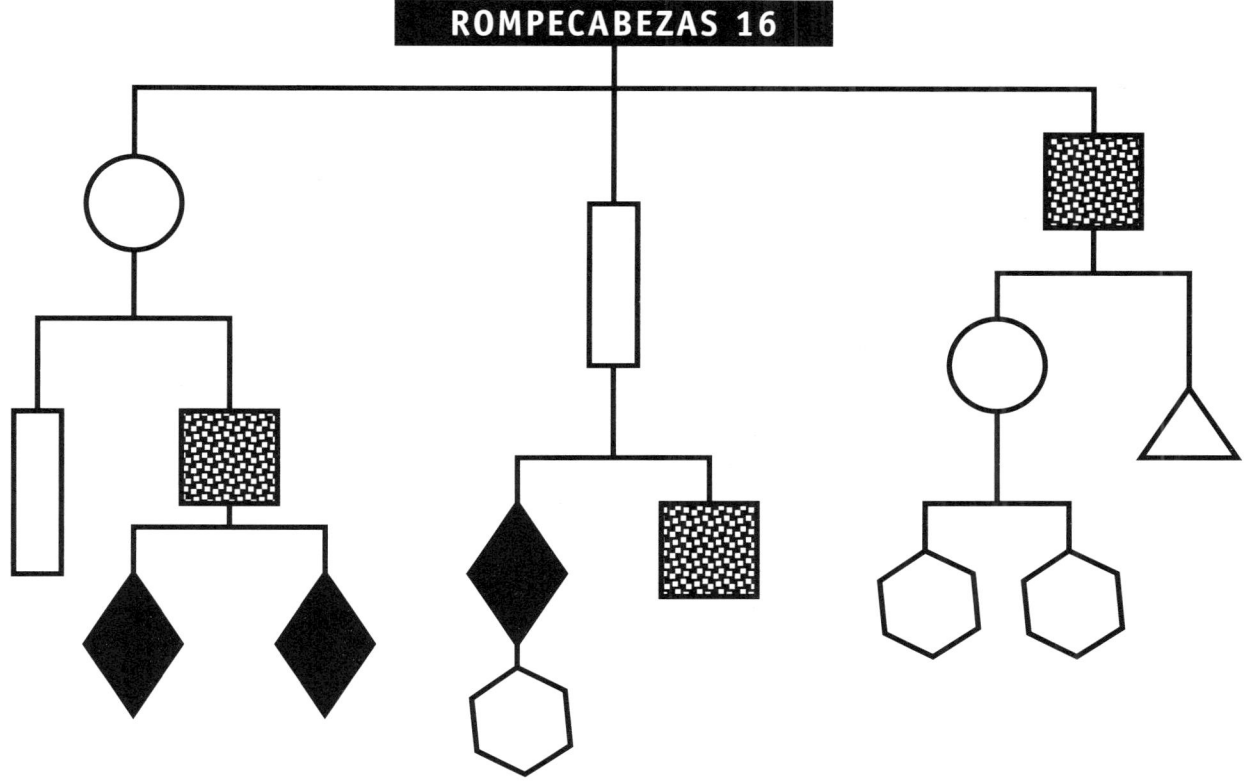

ROMPECABEZAS 16

Encuentra el valor de cada forma. El valor total
es 57. Cada rama tiene el mismo valor. Pista
adicional:

☐ **es un múltiplo de tres.**

69

Nombre _____

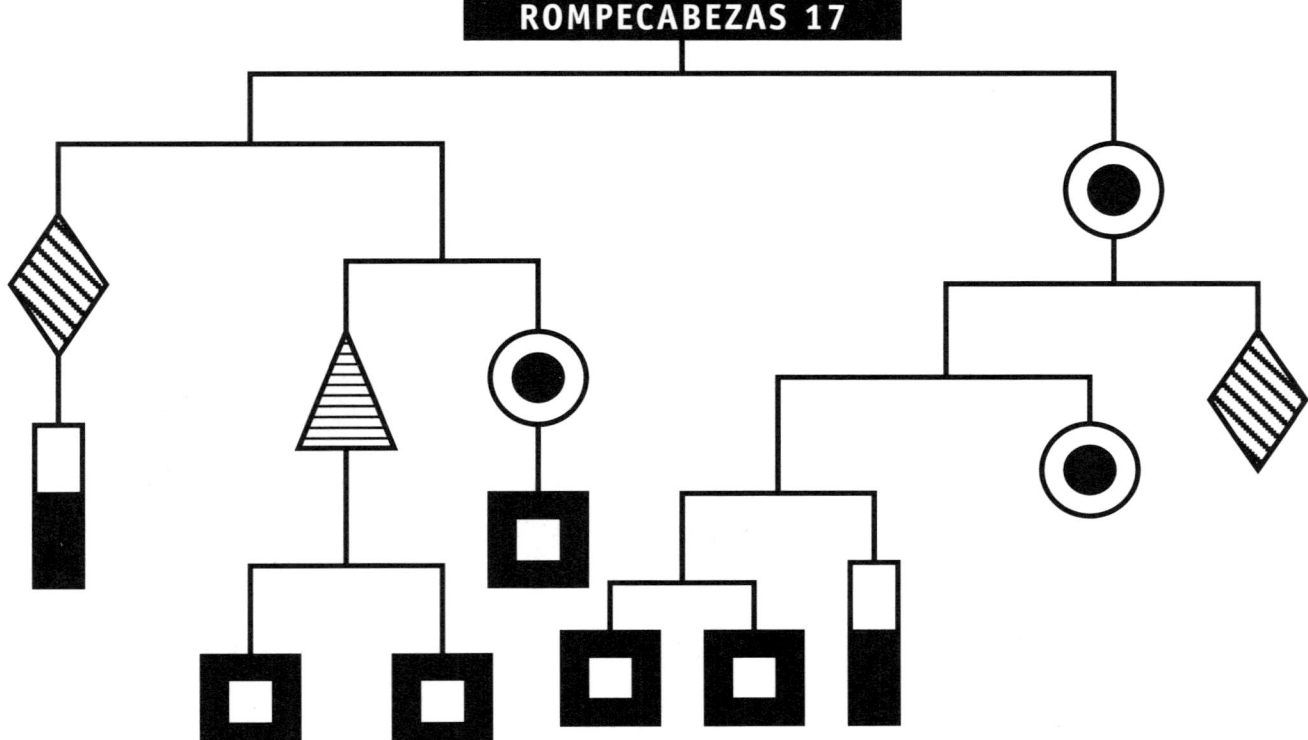

Encuentra el valor de cada forma. El valor total
es 120. Pista:

Todas las formas son múltiplos de ■.

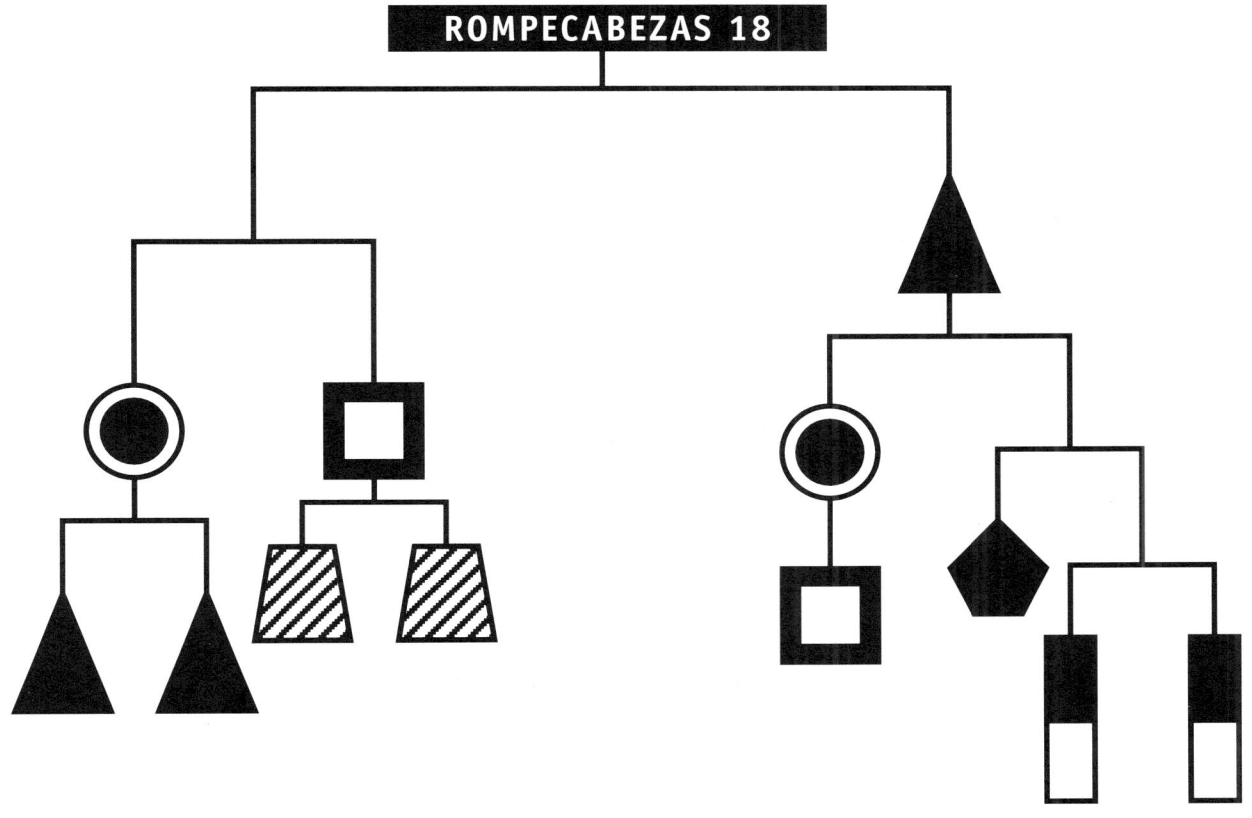

ROMPECABEZAS 18

Encuentra el valor de cada forma. El valor total es 84. Hay tres valores pares y tres valores impares.

Nombre _____

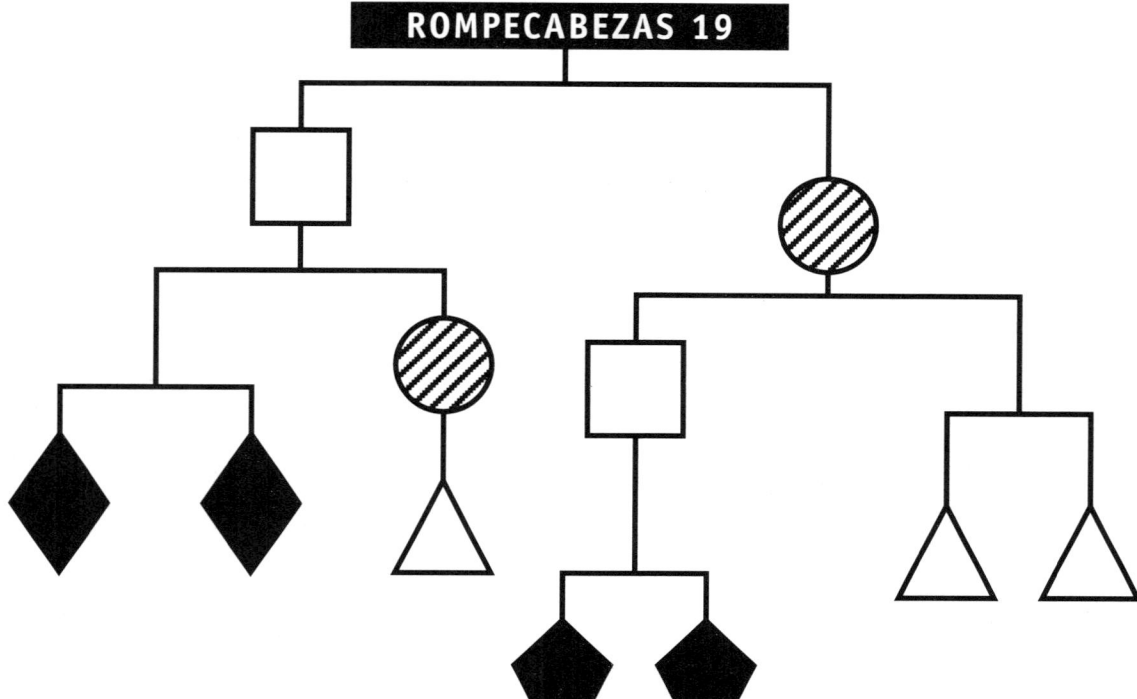

Encuentra el valor de cada forma. El valor total
es 68. Pista:

☐ + ◆ = ⊘

IN THE BALANCE Grades 4–6

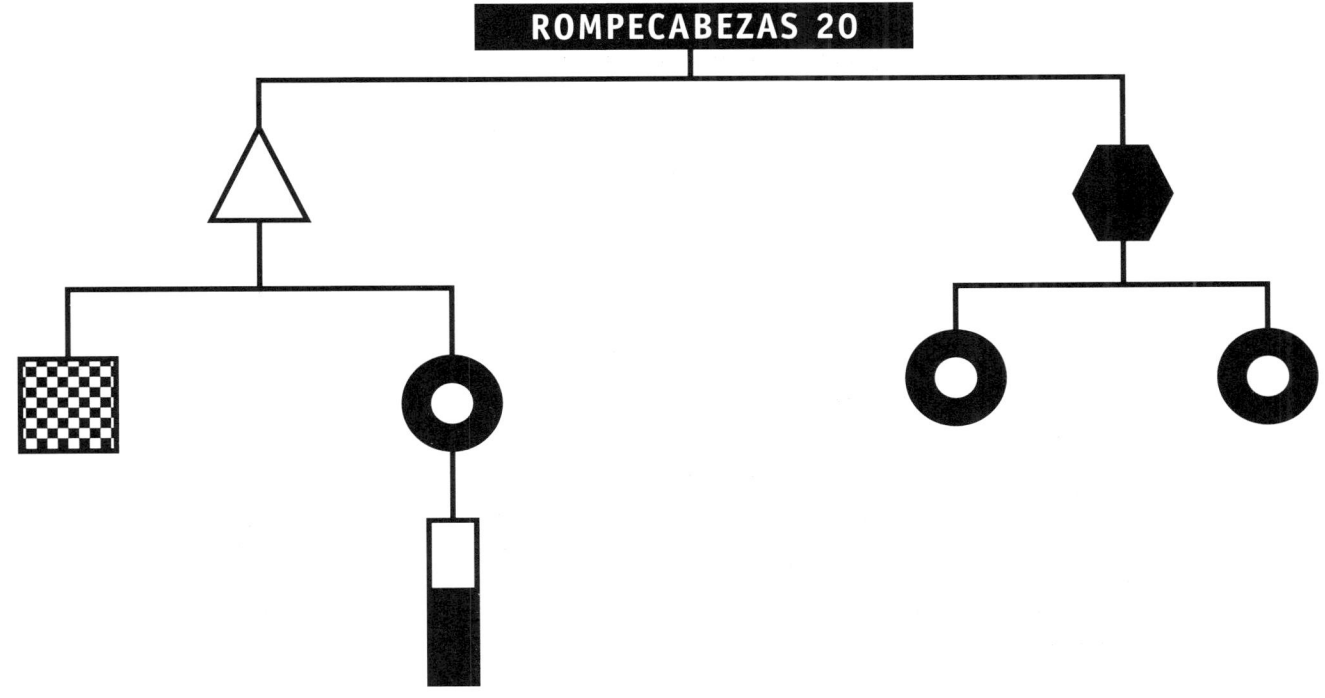

Encuentra el valor de cada forma. El valor total
es 30. Pistas:

3 ▨ < ⬡   △ − ▨ = ▯

©Crective Publications®   **73**

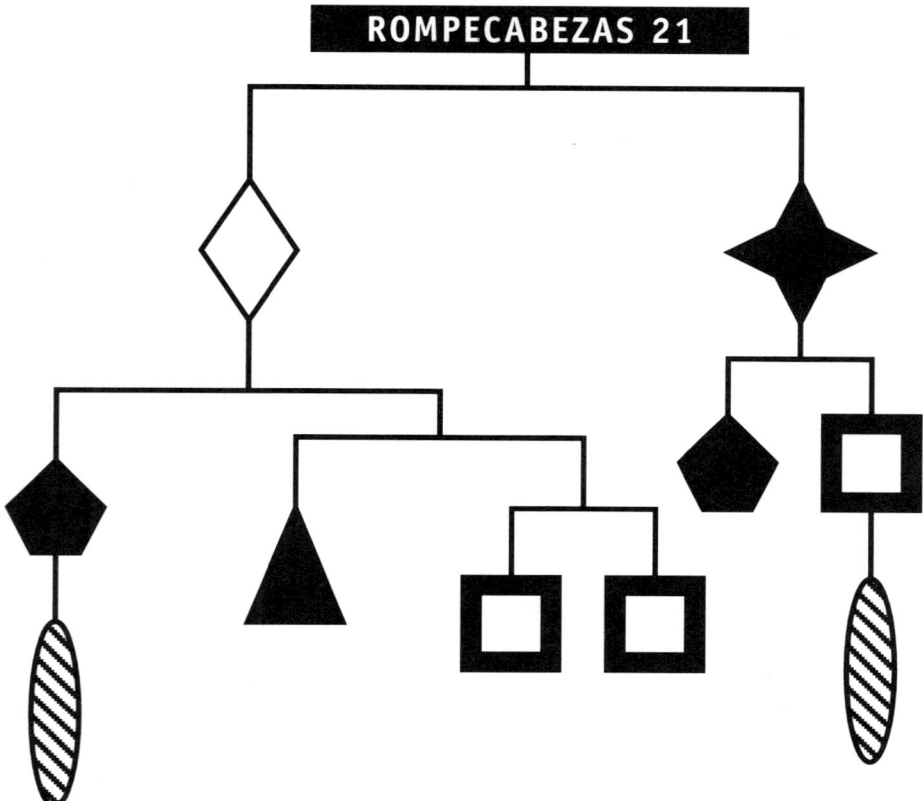

Encuentra el valor de cada forma. El valor total
es 62. Pista:

◇ – ◼ **es un número primo.**

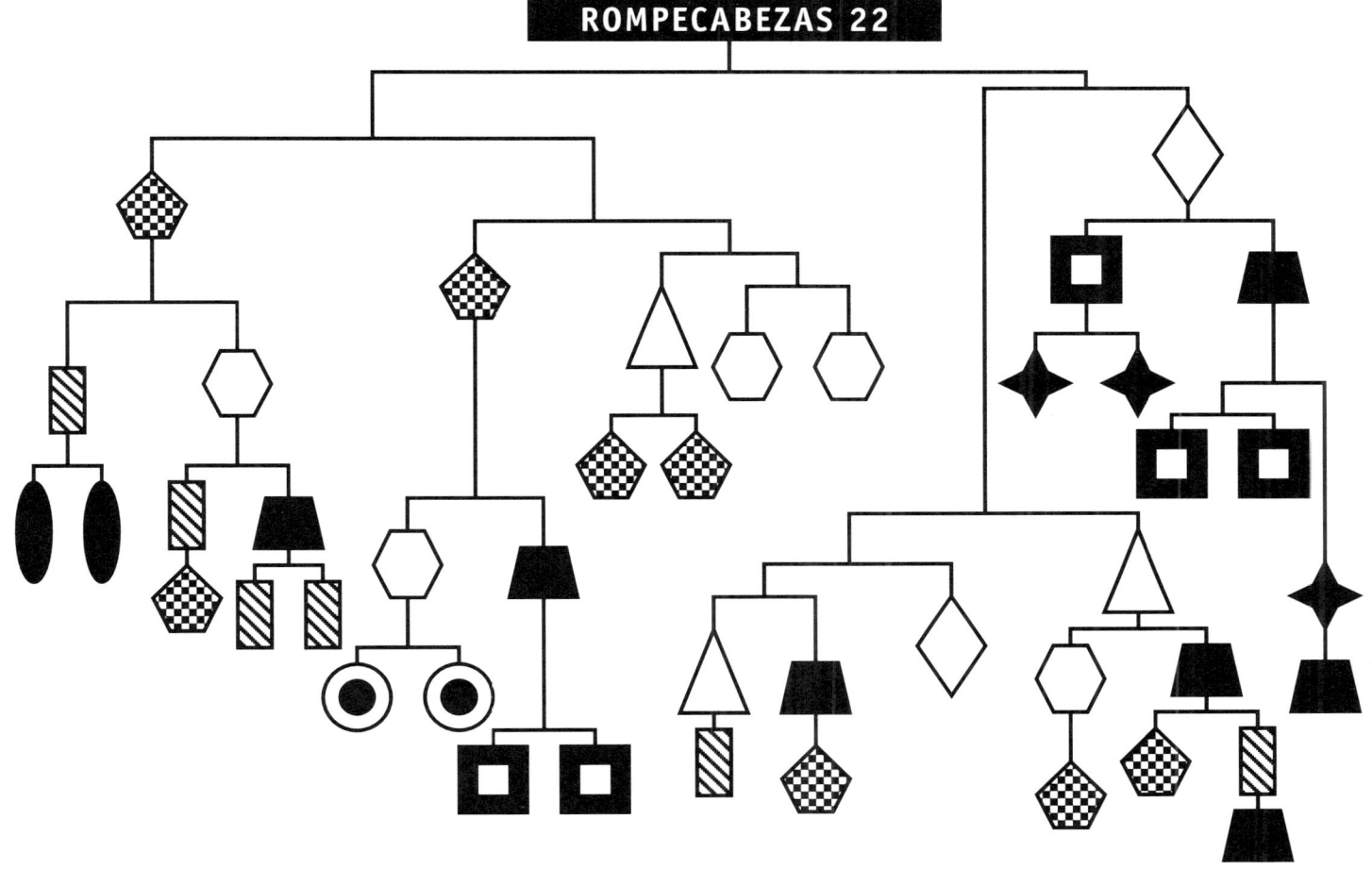

ROMPECABEZAS 22

Encuentra el valor de cada forma.
El valor total es 416. Pista:

▨ − △ = 1

ROMPECABEZAS 23

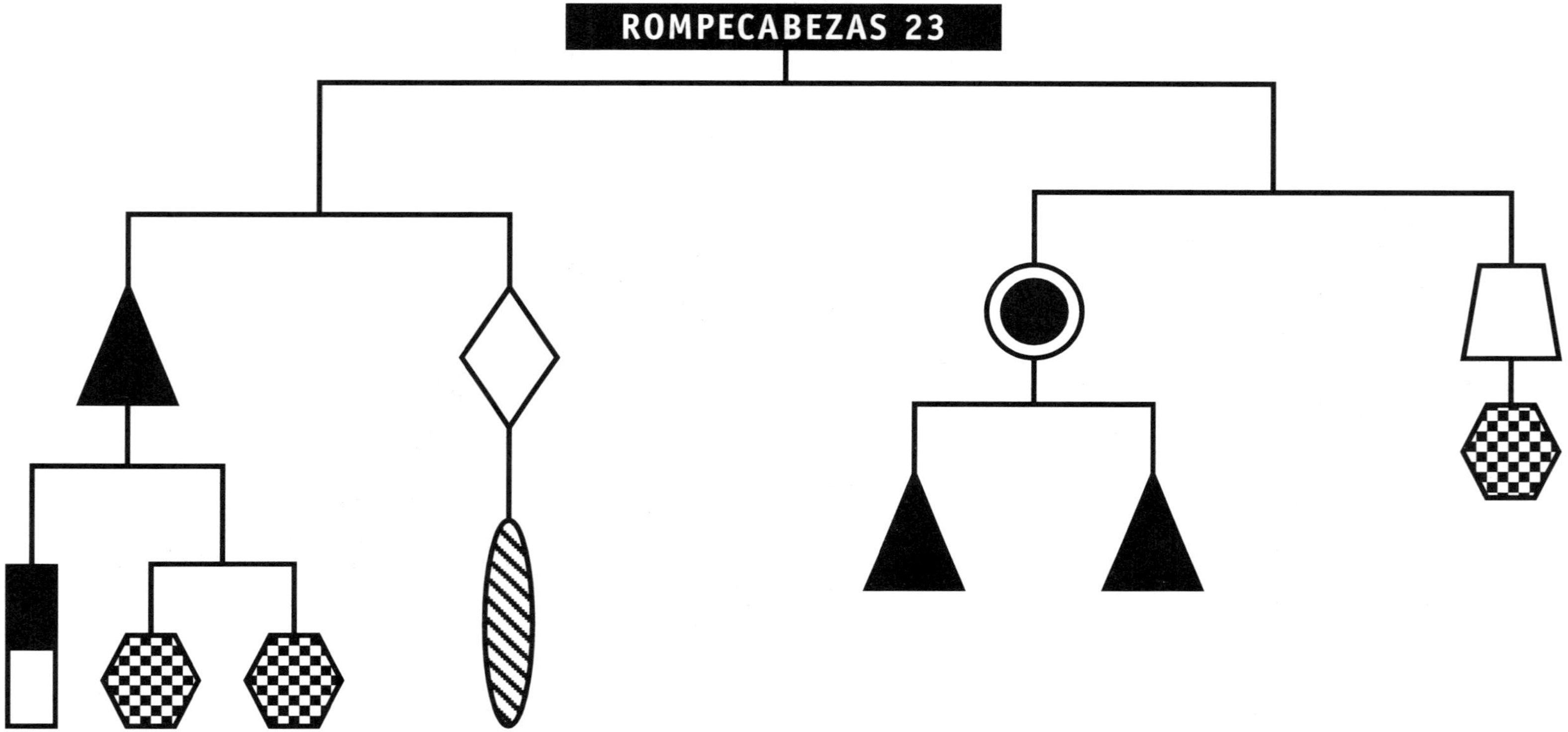

Encuentra el valor de cada forma.
El valor total es 44. Pista:

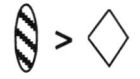 > ◇

76

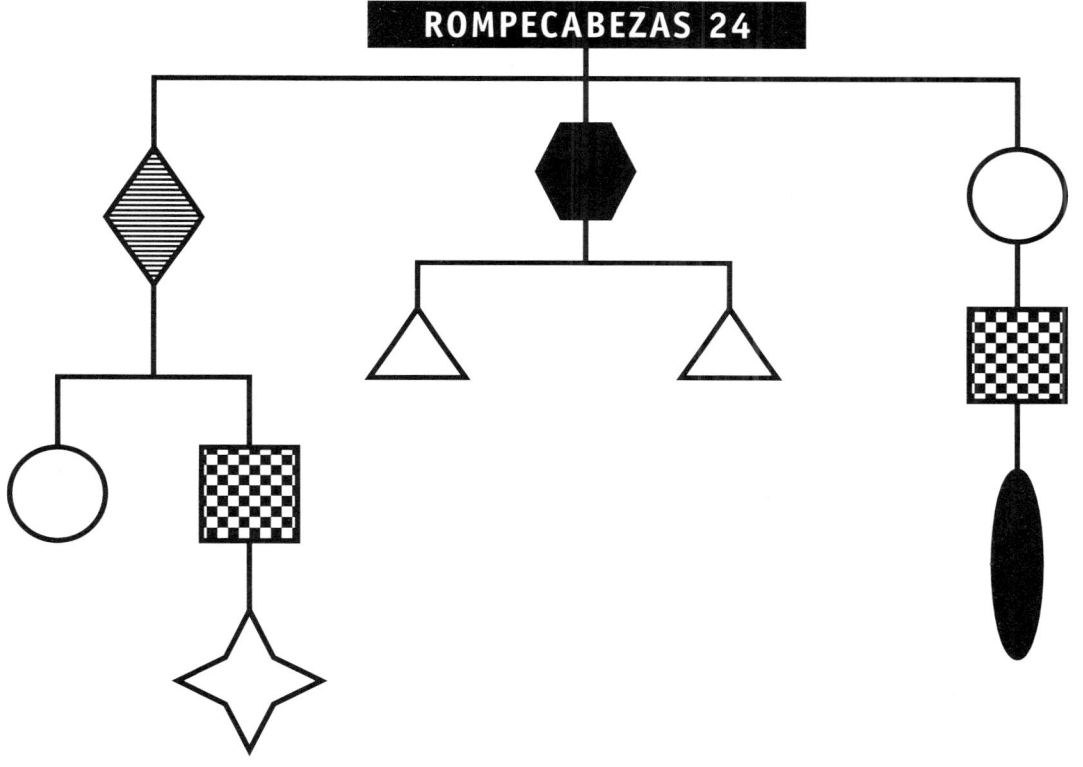

Encuentra el valor de cada forma. El valor total es 51. Cada rama tiene el mismo valor. Pistas adicionales:

 **tienen valores numéricos consecutivos.**

© Crective Publications®

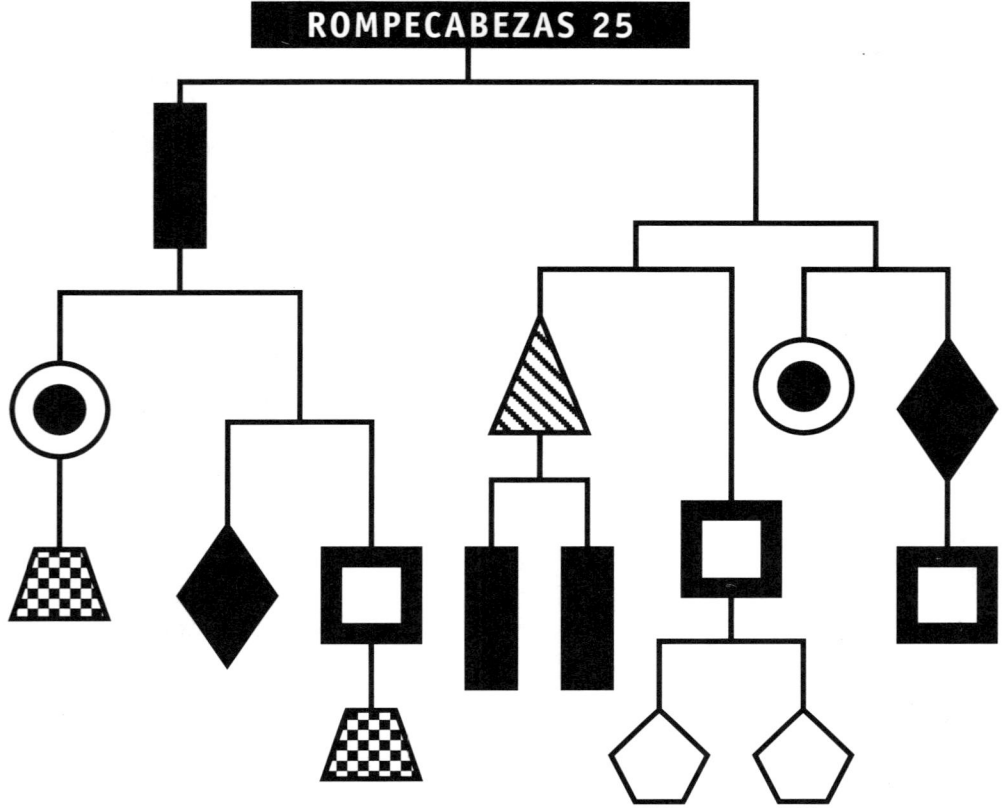

ROMPECABEZAS 25

Encuentra el valor de cada forma.
El valor total es 88. Pista:

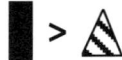 > △